DARK

DATA

DARK DATA

WHY WHAT YOU DON'T KNOW MATTERS

DAVID J. HAND

PRINCETON UNIVERSITY PRESS
PRINCETON AND OXFORD

Published by Princeton University Press
41 William Street, Princeton, New Jersey 08540
6 Oxford Street, Woodstock, Oxfordshire OX20 1TR

press.princeton.edu

First paperback printing 2022
Paperback ISBN 9780691234465

The Library of Congress has cataloged the cloth edition as follows:

Names: Hand, D. J. (David J.), 1950–author.
Title: Dark data : why what you don't know matters / David J. Hand.
Description: Princeton : Princeton University Press, [2020] | Includes
bibliographical references and index.
Identifiers: LCCN 2019022971 (print) | LCCN 2019022972 (ebook) |
ISBN 9780691182377 (hardback) | ISBN 9780691198859 (ebook)
Subjects: LCSH: Missing observations (Statistics) | Big data.
Classification: LCC QA276 .H3178 2020 (print) | LCC QA276 (ebook) | DDC
519.5—dc23
LC record available at https://lccn.loc.gov/2019022971
LC ebook record available at https://lccn.loc.gov/2019022972

British Library Cataloging-in-Publication Data is available

Editorial: Ingrid Gnerlich and Arthur Werneck
Production Editorial: Karen Carter
Text Design: Leslie Flis
Jacket/Cover Design: Jason Alejandro
Production: Jacquie Poirier
Publicity: Sara Henning-Stout and Katie Lewis
Copyeditor: Barbara Liguori

This book has been composed in Arno

To Shelley

CONTENTS

PREFACE

This book is unusual. Most books about data—be they popular books about big data, open data, or data science, or technical statistical books about how to analyze data—are about the data you have. They are about the data sitting in folders on your computer, in files on your desk, or as records in your notebook. In contrast, this book is about data you *don't* have—perhaps data you wish you had, or hoped to have, or thought you had, but nonetheless data you don't have. I argue, and illustrate with many examples, that the missing data are at least as important as the data you do have. The data you cannot see have the potential to mislead you, sometimes even with catastrophic consequences, as we shall see. I show how and why this can happen. But I also show how it can be avoided—what you should look for to sidestep such disasters. And then, perhaps surprisingly, once we have seen how dark data arise and can cause such problems, I show how you can use the dark data perspective to flip the conventional way of looking at data analysis on its head: how hiding data can, if you are clever enough, lead to deeper understanding, better decisions, and better choice of actions.

The question of whether the word *data* should be treated as singular or plural has been a fraught one. In the past it was typically treated as plural, but language evolves, and many people now treat it as singular. In this book I have tried to treat "data" as plural except in those instances where to do so sounded ugly to my ears. Since beauty is said to be in the eye of the beholder, it is entirely possible that my perception may not match yours.

My own understanding of dark data grew slowly throughout my career, and I owe a huge debt of gratitude to the many people who brought me challenges which I slowly realized were dark data problems and who worked with me on developing ways to cope with them. These problems ranged over medical research, the pharmaceutical industry, government and social policy, the financial sector, manufacturing, and other domains. No area is free from the risks of dark data.

Particular people who kindly sacrificed their time to read drafts of the book include Christoforos Anagnostopoulos, Neil Channon, Niall Adams, and three anonymous publisher's readers. They prevented me from making too many embarrassing mistakes. Peter Tallack, my agent, has been hugely supportive in helping me find the ideal publisher for this work, as well as graciously advising me and steering the emphasis and direction of the book. My editor at Princeton University Press, Ingrid Gnerlich, has been a wise and valuable guide in helping me beat my draft into shape. Finally, I am especially grateful to my wife, Professor Shelley Channon, for her thoughtful critique of multiple drafts. The book is significantly improved because of her input.

Imperial College, London

DARK DATA

THEIR ORIGINS AND CONSEQUENCES

Chapter 1

DARK DATA

What We Don't See Shapes Our World

The Ghost of Data

First, a joke.

Walking along the road the other day, I came across an elderly man putting small heaps of powder at intervals of about 50 feet down the center of the road. I asked him what he was doing. "It's elephant powder," he said. "They can't stand it, so it keeps them away."

"But there are no elephants here," I said.

"Exactly!" he replied. "It's wonderfully effective."

Now, on to something much more serious.

Measles kills nearly a 100,000 people each year. One in 500 people who get the disease die from complications, and others suffer permanent hearing loss or brain damage. Fortunately, it's rare in the United States; for example, only 99 cases were reported in 1999. But a measles outbreak led Washington to declare a statewide emergency in January 2019, and other states also reported dramatically increased numbers of cases.[1] A similar pattern was reported elsewhere. In Ukraine, an outbreak resulted in over 21,000 cases by mid-February 2019.[2] In Europe there were 25,863 cases in 2017, but in 2018 there were over 82,000.[3] From

1 January 2016 through the end of March 2017, Romania reported more than 4,000 cases and 18 deaths from measles.

Measles is a particularly pernicious disease, spreading undetected because the symptoms do not become apparent until some weeks after you contract it. It slips under the radar, and you have it before you even know that it's around.

But the disease is also preventable. A simple vaccination can immunize you against the risk of contracting measles. And, indeed, national immunization programs of the kind carried out in the United States have been immensely successful—so successful in fact that most parents in countries which carry out such programs have never seen or experienced the terrible consequences of such preventable diseases.

So, when parents are advised to vaccinate their children against a disease they have neither seen nor heard of any of their friends or neighbors having, a disease which the Centers for Disease Control and Prevention announced was no longer endemic in the United States, they naturally take the advice with a pinch of salt.

Vaccinate against something which is not there? It's like using the elephant powder.

Except that, unlike the elephants, the risks are still there, just as real as ever. It's merely that the information and data these parents need to make decisions are missing, so that the risks have become invisible.

My general term for the various kinds of missing data is *dark data*. Dark data are concealed from us, and that very fact means we are at risk of misunderstanding, of drawing incorrect conclusions, and of making poor decisions. In short, our ignorance means we get things wrong.

The term "dark data" arises by analogy with the dark matter of physics. About 27 percent of the universe consists of this mysterious substance, which doesn't interact with light or other electromagnetic radiation and so can't be seen. Since dark matter can't be seen, astronomers were long unaware of its existence. But then observations of the rotations of galaxies revealed that the more distant stars were not moving more slowly than stars nearer the center, contradicting what we would have expected from our understanding of gravity. This rotational anomaly can be explained by supposing that galaxies have more mass than appears to be the case judging from the stars and other objects we can see through our telescopes. Since we can't see this extra mass, it has been called dark matter. And it can be significant (I almost said "it can matter"): our home galaxy, the Milky Way, is estimated to have some ten times as much dark matter as ordinary matter.

Dark data and dark matter behave in an analogous way: we don't see such data, they have not been recorded, and yet they can have a major effect on our conclusions, decisions, and actions. And as some of the later examples will show, unless we are aware of the possibility that there's something unknown lurking out there, the consequences can be disastrous, even fatal.

The aim of this book is to explore just how and why dark data arise. We shall look at the different kinds of dark data and see what leads to them. We shall see what steps we can take to avoid dark data's arising in the first place. We shall see what we can do when we realize that dark data are obscured from us. Ultimately, we shall also see that if we are clever enough, we can sometimes take advantage of dark data. Curious and paradoxical though that may seem, we can make use of ignorance and the dark data perspective to enable better decisions and take better actions. In practical terms, this means we can lead healthier lives, make more money, and take lower risks by judicious use of the unknown. This doesn't mean we should hide information from others (though, as we shall also see, deliberately concealed data is one common kind of dark data). It is much more subtle than that, and it means that everyone can benefit.

Dark data arise in many different shapes and forms as well as for many different reasons, and this book introduces a taxonomy of such reasons, the *types* of dark data, labeled *DD-Type x*, for "Dark Data-Type x." There are 15 *DD-Types* in all. My taxonomy is not exhaustive. Given the wealth of reasons for dark data, that would probably be impossible. Moreover, any particular example of dark data might well illustrate the effect of more than one *DD-Type* simultaneously—*DD-Types* can work together and can even combine in an unfortunate synergy. Nonetheless, an awareness of these *DD-Types*, and examination of examples showing how dark data can manifest, can equip you to identify when problems occur and protect you against their dangers. I list the *DD-Types* at the end of this chapter, ordered roughly according to similarity, and describe them in more detail in chapter 10. Throughout the book I have indicated some of the places when an example of a particular *Type* occurs. However, I have deliberately not tried to do this in an exhaustive way—that would be rather intrusive.

To get us going, let's take a new example.

In medicine, trauma is serious injury with possible major long-term consequences. It's one of the most serious causes of "life years lost" through premature death and disability, and is the commonest cause of death for those under age 40. The database of the Trauma Audit and Research Network (TARN) is the largest medical trauma database in Europe. It receives data on trauma events from more than 200 hospitals, including over 93 percent of the hospitals in England and Wales, as well as hospitals in Ireland, the Netherlands, and Switzerland. It's clearly a very rich seam of data for studying prognoses and the effectiveness of interventions in trauma cases.

Dr. Evgeny Mirkes and his colleagues from the University of Leicester in the UK looked at some of the data from this database.[4] Among the 165,559 trauma cases they examined, they found 19,289 with unknown outcomes. "Outcome" in trauma research means whether or not the patient survives at least 30 days after the injury. So the 30-day survival was unknown for over 11 percent of the patients. This example illustrates a common form of dark data—our *DD-Type 1: Data We Know Are Missing*. We know these patients had some outcome—we just don't know what it was.

No problem, you might think—let's just analyze the 146,270 patients for whom we do know the outcome and base our understanding and prognoses on those. After all, 146,270 is a big number—within the realm of medicine it's "big data"—so surely we can be confident that any conclusions based on these data will be right.

But can we? Perhaps the missing 19,289 cases are very different from the others. After all, they were certainly different in that they had unknown outcomes, so it wouldn't be unreasonable to suspect they might differ in other ways. Consequently, any analysis of the 146,270 patients with known outcomes might be misleading relative to the overall population of trauma patients. Thus, actions taken on the basis of such analysis might be the wrong actions, perhaps leading to mistaken prognoses, incorrect prescriptions, and inappropriate treatment regimes, with unfortunate, even fatal, consequences for patients.

To take a deliberately unrealistic and extreme illustration, suppose that all 146,270 of those with known outcomes survived and recovered without treatment, but the 19,289 with unknown outcomes all died within two days of admission. If we ignored those with unknown outcomes, we would justifiably conclude there was nothing to worry about, and all patients with trauma recovered. On this basis, we wouldn't treat any incoming trauma cases, expecting them to recover naturally. And then we would be horrified and confused by the fact that more than 11 percent of our patients were dying.

Before I go any further with this story, I want to reassure the reader. My extreme illustration is very much a worst-case scenario—we might reasonably expect things not to be that bad in reality—and Dr. Mirkes and his colleagues are experts on missing data analysis. They are very aware of the dangers and have been developing statistical methods to cope with the problem; I describe similar methods later in this book. But the take-home message from this story is that *things may not be what they seem*. Indeed, if there were a single take-home message from this book, that would be a good approximation to it: while it helps to have lots of data—that is, "big data"—size is not everything. And what you don't know, the data you don't have, may be even more important in understanding what's going on than the data you do have. In any case, as we shall see, the problems of dark data are not merely big-data problems: they also arise with small data sets. They are ubiquitous.

My story about the TARN database may be exaggerated, but it serves as a warning. Perhaps the outcomes of the 19,289 patients were not recorded precisely *because* they'd all died within 30 days. After all, if the outcome was based on contacting the patients 30 days after admission to see how they were, none of those who died would respond to questions. Unless we were aware of this possibility, we'd never record that any patients had died.

This may sound a bit silly, but in fact such situations arise quite often. For example, a model built to determine the prognosis for patients being given a particular treatment will be based on the outcomes of previous patients who received that treatment. But what if insufficient time had passed for all such previous patients to have reached an outcome? For those patients the eventual outcome would be unknown. A model built just on those for whom the outcome was known could be misleading.

A similar phenomenon happens with surveys, in which *nonresponse* is a source of difficulty. Researchers will typically have a complete list of people from whom they would ideally like answers, but, also typically, not everyone responds. If those who do respond differ in some way from those who do not, then the researchers might have cause to doubt whether the statistics are good summaries of the population. After all, if a magazine carried out a survey of its subscribers asking the single question, Do you reply to magazine surveys? then we could not interpret the fact that 100 percent of those who replied answered yes as meaning that all the subscribers replied to such surveys.

The preceding examples illustrate our first type of dark data. We know that the data for the TARN patients all exist, even if the values aren't all recorded. We know that the people on the survey list had answers, even if they did not give them. In general, we know that there are values for the data; we just don't know what those values are.

An illustration of a different kind of dark data (*DD-Type 2: Data We Don't Know Are Missing*) is the following.

Many cities have problems with potholes in road surfaces. Water gets into small cracks and freezes in the winter, expanding the cracks, which are then further damaged by car tires. This results in a vicious circle, ending with a tire- and axle-wrecking hole in the road. The city of Boston decided to tackle this problem using modern technology. It released a smartphone app which used the internal accelerometer of the phone to detect the jolt of a car being driven over a pothole and then used GPS to automatically transmit the location of the hole to the city authorities.

Wonderful! Now the highway maintenance people would know exactly where to go to repair the potholes.

Again, this looks like an elegant and cheap solution to a real problem, built on modern data analytic technology—except for the fact that ownership of cars and expensive smartphones is more likely to be concentrated in wealthier areas. Thus, it's quite likely that potholes in poorer areas would not be detected, so that their location would not be transmitted, and some areas might never have their potholes fixed. Rather than solving the pothole problem in general, this approach might even aggravate social inequalities. The situation here is different from that in the TARN example, in which we knew that certain data were missing. Here we are unaware of them.

The following is another illustration of this kind of dark data. In late October 2012, Hurricane Sandy, also called "Superstorm Sandy,"[5] struck the Eastern Seaboard of the United States. At the time it was the second most costly hurricane in U.S. history and the largest Atlantic hurricane on record, causing damage estimated at $75 billion, and killing more than 200 people in eight countries. Sandy affected 24 U.S. states, from Florida to Maine to Michigan to Wisconsin, and led to the closure of the financial markets owing to power cuts. And it resulted, indirectly, in a surge in the birth rate some nine months later.

It was also a triumph of modern media. The physical storm Hurricane Sandy was accompanied by a Twitter storm of messages describing what was going on. The point about Twitter is that it tells you what and where something is happening as it happens, as well as who it's happening to. The social media platform is a way to keep up in real time as events unfold. And that's exactly what occurred with Hurricane Sandy. Between 27 October and 1 November 2012, there were more than 20 million tweets about it. Clearly, then, we might think, this is ideal material from which to get a continuously evolving picture of the storm as it develops, identifying which areas have been most seriously affected, and where emergency relief is needed.

But later analysis revealed that the largest number of tweets about Sandy came from Manhattan, with few tweets coming from areas like Rockaway and Coney Island. Did that mean that Rockaway and Coney Island were less severely affected? Now it's true that subways and streets of Manhattan were flooded, but it was hardly the worst-hit region, even of New York. The truth is, of course, that those regions transmitting fewer tweets may have been doing so not because the storm had less impact but simply because there were fewer Twitter users with fewer smartphones to tweet them.

In fact, we can again imagine an extreme of this situation. Had any community been completely obliterated by Sandy, then no tweets at all would have emerged. The superficial impression would be that everybody there was fine. Dark data indeed.

As with the first type of dark data, examples of this second kind, in which we don't know that something is missing, are ubiquitous. Think of undetected fraud, or the failure of a crime-victim survey to identify that any murders have been committed.

You might have a sense of déjà vu about those first two types of dark data. In a famous news briefing, former U.S. Secretary of Defense Donald Rumsfeld nicely characterized them in a punchy sound bite, saying "there are known unknowns; that is to say we know there are some things we do not know. But there are also unknown unknowns—the ones we don't know we don't know."[6] Rumsfeld attracted considerable media ridicule for that convoluted statement, but the criticism was unfair. What he said made very good sense and was certainly true.

But those first two types are just the beginning. In the next section we introduce some of the other types of dark data. These, and others described later, are what this book is all about. As you will see, dark data have many forms. Unless we are aware that data might be incomplete, that observing something does not mean observing everything, that a measurement procedure might be inaccurate, and that what is measured might not really be what we want to measure, then we could get a very misleading impression of what's going on. Just because there's no one around to hear that tree fall in the forest doesn't mean that it didn't make a noise.

So You Think You Have All the Data?

The customer arrives at the supermarket checkout with a full shopping cart. The laser scans the barcode of each item, and the till emits its electronic beep as it adds up the total cost. At the end of this exercise, the customer is presented with the overall bill and pays. Except that's not really the end. The data describing the items bought and the price of each are sent to a database and stored. Later, statisticians and data scientists will pore over the data, extracting a picture of customer behavior from details of what items were bought, which items were bought together, and indeed what sort of customer bought the items. Surely there's no opportunity for missing data here? Data of the transaction have to be captured if the supermarket is to work out how much to charge the customer—short of a power cut, register failure, or fraud, that is.

Now it seems pretty obvious that the data collected are all the data there are. It's not just *some* of the transactions or details of just *some* of the items purchased. It's *all* the transactions made by *all* the customers on *all* the items in that supermarket. It is, as is sometimes simply said, "data = all."

But is it really? After all, these data describe what happened *last* week or *last* month. That's useful, but if we are running the supermarket, what we probably really want to know is what will happen tomorrow or next week or next month. We really want to know who will buy what when, and how much of it they will buy in the future. What's likely to run out if we don't put more on the shelves? What brands will people prefer to buy? We really want data that have not been measured. Dark data *DD-Type 7: Changes with Time* describes the obscuring nature of time on data.

Indeed, beyond that complication, we might want to know how people *would have behaved* had we stocked different items, or arranged them differently on the shelves, or changed the supermarket opening times. These are called *counterfactuals* because they are contrary to fact—they are about what would have happened if what actually happened hadn't. Counterfactuals are dark data *DD-Type 6: Data Which Might Have Been.*

Needless to say, counterfactuals are of concern not just to supermarket managers. You've taken medicines in the past. You trusted the doctor who prescribed them, and you assumed they'd been tested and found to be effective in alleviating a condition. But how would you feel if you discovered that they hadn't been tested? That no data had been collected on whether the medicines made things better? Indeed, that it was possible they made things worse? Or that even if they had been tested and found to help, the medicines hadn't been compared with simply leaving the condition alone, to see if they made it get better more quickly than natural healing processes? Or the medicines hadn't been compared with other ones, to see if they were more effective than familiar alternatives? In the elephant powder example, a comparison with doing nothing would soon reveal that *doing nothing was just as effective at keeping the elephants away* as putting down the heaps of powder. (And that, in turn could lead to the observation that there were actually no elephants to be kept away.)

Returning to the notion of "data=all," in other contexts the notion that we might have "all" the data is *clearly* nonsensical. Consider your weight. This is easy enough to measure—just hop on your bathroom scale. But if you repeat the measurement, even very soon afterward, you might find a slightly different result, especially if you try to measure it to the nearest ounce or gram. All physical measurements are subject to potential inaccuracies as a result of measurement error or random fluctuations arising from very slight changes in the circumstances (*DD-Type 10: Measurement Error and Uncertainty*). To get around this problem, scientists measuring the magnitude of some phenomenon—the speed of light, say, or the electric charge of the electron—will take multiple measurements and average them. They might take 10 measurements, or 100. But what they obviously cannot do is take "all" the measurements. There is no such thing as "all" in this context.

A different type of dark data is illustrated when you ride on London's red buses: you will know that more often than not they are packed with passengers. And yet data show that the occupancy of the average bus is just 17 people. What can explain this apparent contradiction? Is someone manipulating the figures?

A little thought reveals that the answer is simply that more people are riding on the buses when they are full—that's what "full" means. The consequence is that more people see a full bus. At the opposite extreme, an empty bus will have no one to report that it was empty. (I'm ignoring the driver in all this, of course.) This example is an illustration of dark data *DD-Type 3: Choosing Just Some Cases*. Furthermore, that mode of dark data can even be a necessary consequence of collecting data, in which case it illustrates *DD-Type 4: Self-Selection*. The following are my two favorite examples of opposite extremes in terms of significance.

The first is the cartoon showing a man looking at one of those maps which are placed outside railway stations. In the middle of the map is a red dot with a label saying "You are here." "How," thinks the man, "did they know?" They knew because they recognized that *everyone* looking at the red dot had to be in front of the sign. It was a highly selected sample and *necessarily* missed everyone standing elsewhere.

The point is that data can be collected only if there is someone or something—a measuring instrument, for example—there to collect them. And the second extreme manifestation of this is described by the *anthropic principle*, which essentially says that the universe has to be like it is, or we would not be here to observe it. We cannot have data from very different universes because we could not exist in those and so could not collect data from them. This means any conclusions we draw are necessarily limited to our (type of) universe: as with the potholes, there might be all sorts of other things going on which we don't know about.

There's an important lesson for science here. Your theory might be perfectly sound for your data, but your data will have limits. They might not refer to very high temperatures or long times or vast distances. And if you extrapolate beyond the limits within which your data were collected, then perhaps your theory will break down. Economic theories built on data collected during benign conditions can fail dramatically in recessions, and Newton's laws work fine unless tiny objects or high velocities or other extremes are involved. This is the essence of *DD-Type 15: Extrapolating beyond Your Data.*

I have a T-shirt with an *xkcd* cartoon with two characters talking to each other. One character says "I used to think correlation implied causation." In the next frame, he goes on to say, "Then I took a statistics class. Now I don't." Finally, the other character says, "Sounds like the class helped," and the first character replies, "Well, maybe."[7]

Correlation simply means that two things vary together: for example, positive correlation means that when one is big then the other is big, and when the first is small, the second is small. That's different from causation. One thing is said to *cause* another if a change in the first induces a change in the second. And the trouble is that two things can vary together without changes in one being the cause of changes in the other. For example, observations over the early years of schooling show that children with a larger vocabulary tend, on average, to be taller. But you wouldn't then believe that parents who wanted taller offspring should hire tutors to expand their vocabulary. It's more likely that there are some unmeasured dark data, a third factor which accounts for the correlation—such as the ages of the children. When the *xkcd* character says, "Well, maybe," he's acknowledging that it's possible that taking the statistics class caused his understanding to change, but maybe there was some other cause. We shall see some striking examples of this situation, characterized by *DD-Type 5: Missing What Matters.*

I've now mentioned several dark data types. But there are more. The aim of this book is to reveal them, to show how they can be identified, to observe their impact, and to show how to tackle the problems they cause—and even how to take advantage of them. They are listed at the end of this chapter, and their content is summarized in chapter 10.

Nothing Happened, So We Ignored It

A final example illustrates that dark data can have disastrous consequences and that they are not especially a problem of large data sets.

Thirty years ago, on 28 January 1986, 73 seconds into its flight and at an altitude of 9 miles, the space shuttle *Challenger* experienced an enormous fireball caused by one of its two booster rockets and broke up. The crew compartment continued its trajectory, reaching an altitude of 12 miles, before falling into the Atlantic. All seven crew members, consisting of five astronauts and two payload specialists, were killed.

A later presidential commission found that NASA middle managers had violated safety rules requiring data to be passed up the chain of command. This was attributed to economic pressures, making it very important that the launch schedule should be maintained: the launch date had already slipped from January 22nd to the 23rd, then to the 25th, and then to the 26th. Since temperature forecasts for that day suggested an unacceptably low temperature, the launch was again rescheduled, for the 27th. Countdown proceeded normally until indicators suggested a hatch lock had not closed properly. By the time that was fixed the wind was too strong, and again the launch was postponed.

On the night of January 27th, a three-hour teleconference was held between Morton Thiokol, which was the company that made the booster rockets, NASA staff at the Marshall Space Flight Center, and people from the Kennedy Space Center. Larry Wear, of the Marshall Center, asked Morton Thiokol to check the possible impact of low temperatures on the solid rocket motors. In response, the Morton Thiokol team pointed out that the O-rings would harden in low temperatures.

The O-rings were rubber-like seals, with a cross-section diameter of about a quarter of an inch, which fitted in the joint around the circumference between each of the four rocket motor segments. The solid rocket boosters were 149 feet high and 38 feet in circumference. Under launch conditions, the 0.004 inch gap that the O-rings normally sealed typically opened to a maximum of 0.06 inch: just six one-hundredths of an inch. And during launch this larger gap remained open for just six-tenths of a second.

Robert Ebeling of Morton Thiokol had been concerned that at low temperatures the hardening of the O-rings meant they would lose their ability to create an effective seal between segments when the gaps expanded by that 0.056 inch for that 0.6 second. At the teleconference Robert Lund, vice president of Morton Thiokol, said that the O-ring operating temperature must not be less than the previous lowest launch temperature, 53°F. Extensive, sometimes heated, discussion ensued, both in the conference and off-line in private conversations. Eventually, Morton Thiokol reconsidered and recommended launch.

Precisely 58.79 seconds after the launch a flame burst from the right solid rocket motor near the last joint. This flame quickly grew into a jet which broke the struts joining the solid rocket motor to the external fuel tank. The motor pivoted, hitting first the Orbiter's wing and then the external fuel tank. The jet of flame then fell onto this external tank containing the liquid hydrogen and oxygen fuel. At 64.66 seconds the tank's surface was breached, and 9 seconds later *Challenger* was engulfed in a ball of flame and broke into several large sections.[8]

One thing we have to remember is that space travel is all about risk. No mission, even under the very best of circumstances, is a risk-free enterprise: the risk cannot be reduced to zero. And there are always competing demands.

Furthermore, as with any incident like this, the notion of "cause" is complicated. Was it due to violation of safety rules, undue pressure put on managers because of economic considerations, other consequences of budget tightening, or perhaps media pressure following the fact that the launch of the previous shuttle, *Columbia*, had been delayed seven times, each delay greeted with press ridicule? For example, here's Dan Rather's script for the evening news on Monday, January 27th, following the four delays to the *Challenger* launch: "Yet another costly, red-faces-all-around space-shuttle-launch delay. This time a bad bolt on a hatch and a bad-weather bolt from the blue are being blamed." Or was it a consequence of political pressure. After all, there was significantly more interest in this launch than earlier launches because it carried an "ordinary person," Christa McAuliffe, a teacher, and the president's State of the Union address was scheduled for the evening of January 28th.

In such situations, multiple factors typically come together. Complex and obscure interactions can lead to unexpected consequences. But in this case there was another factor: dark data.

After the disaster, a commission headed by former secretary of state William Rogers drew attention to the fact that flights which had not had any O-rings showing distress had not been included in the diagram discussed at the teleconference (dark data *DD-Type 3: Choosing Just Some Cases* but also *DD-Type 2: Data We Don't Know Are Missing*). The report said (p. 146): "The managers compared as a function of temperature the flights for which thermal distress of O-rings had been observed—not the frequency of occurrence based on all flights."[9] And that's the give-away: *data from some flights were not included in the analysis*. My earlier examples have shown the sorts of problems leaving out some of the data can lead to.

The report went on: "In such a comparison [that is, using the limited set of data presented], there is nothing irregular in the distribution of O-ring 'distress' over the spectrum of joint temperatures at launch between 53 degrees Fahrenheit and 75 degrees Fahrenheit," meaning: there is no apparent relationship between temperature and number of O-rings showing distress. However, "when the entire history of flight experience is considered, including 'normal' flights with no erosion or blow-by, the comparison is substantially different"; that is, if you include all the data, you get a different picture. In fact, flights which took place at higher temperatures were much more likely to show no problems, and these were the dark data not shown in the plot. But if the higher the temperature, the less the chance of a problem, then, conversely, the lower the temperature, the greater the chance of a problem. And the ambient temperature was predicted to be just 31°F.

This section of the report concluded: "Consideration of the entire launch temperature history indicates that the probability of O-ring distress is increased to *almost a certainty* if the temperature of the joint is less than 65[°F]." (my italics)

The situation is graphically illustrated in the two diagrams in Figure 1. Figure 1(a) shows the diagram discussed at the teleconference. This is a plot of the number of distressed O-rings on each launch plotted against launch temperature in degrees Fahrenheit. So, for example, at the lowest launch temperature in the past, 53°F, three of the O-rings experienced distress, and at the highest launch temperature in the past, 75°F, two of the O-rings experienced distress. There is no clear relationship between launch temperature and the number of distressed O-rings.

However, if we add the missing data—showing the launches which led to no O-ring distress, we obtain Figure 1(b). The pattern is now very clear. In fact, *all* the launches which occurred

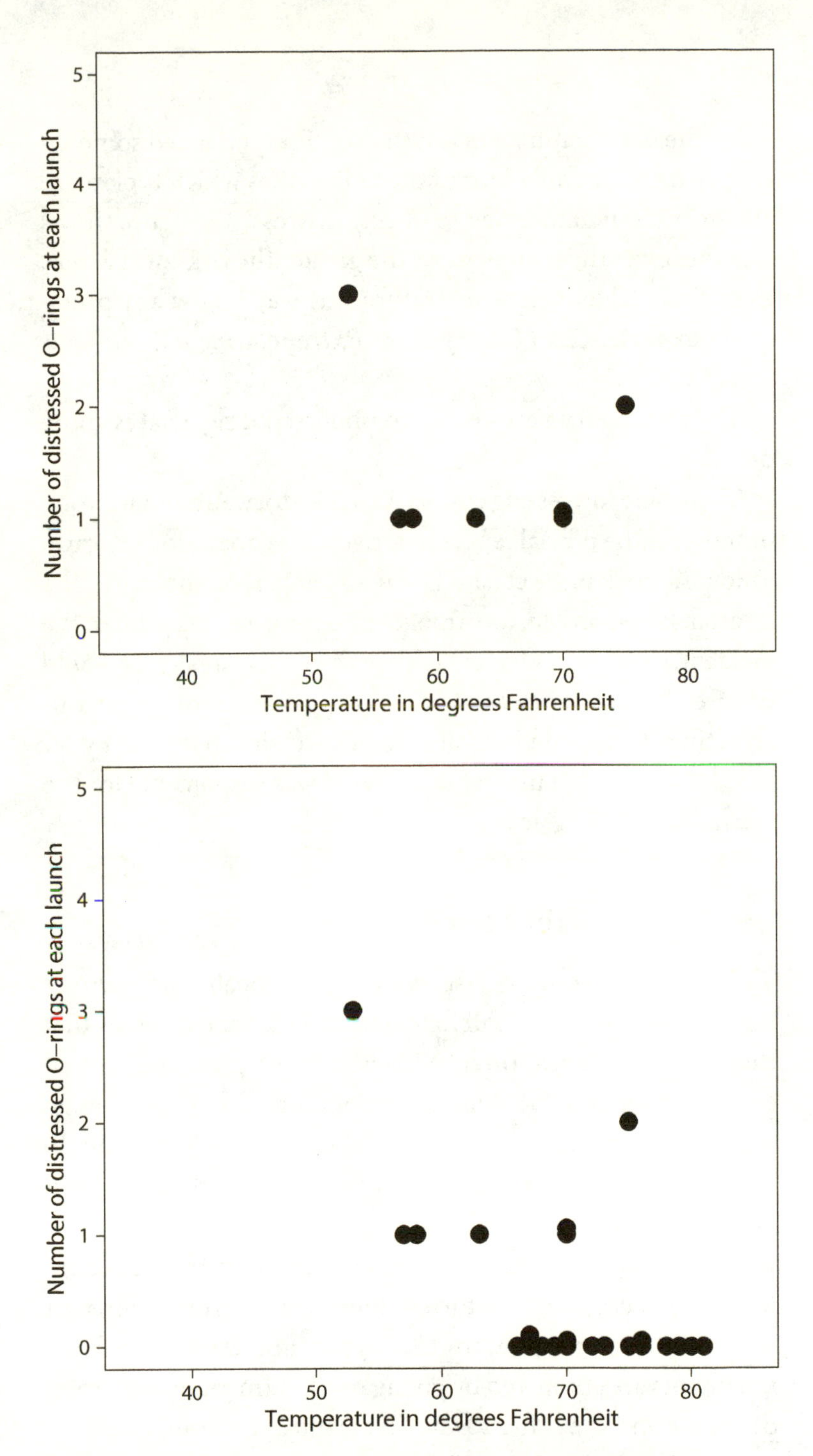

FIGURE 1. (a) Data examined in the *Challenger* prelaunch teleconference; (b) complete data.

when the temperature was less than 65°F experienced some O-ring distress, but only four of the 21 launches which took place at higher temperatures led to O-ring distress. The figure shows that the lower the temperature, the greater the risk. And worse, the predicted launch temperature was way below any previously experienced (*DD-Type 15: Extrapolating beyond Your Data*).

The missing data are crucial to understanding what is going on.

There is an interesting corollary to this story. Although it took months for the official report to arrive at its conclusion, Morton Thiokol's stock price crashed by 11.86 percent on the day of the disaster. Based on Morton Thiokol price movements prior to the incident, stock price changes of even as much as 4 percent would be rare. The stock prices of other companies involved in constructing the shuttle launch vehicle also slumped, but by far less. It was as if the market knew what was responsible for the crash. Dark data again?

The Power of Dark Data

That last example shows just how catastrophically badly wrong things can go when we fail to allow for dark data. It seems that they represent a real threat. But in fact the picture is not all gloomy. It turns out that an understanding of dark data can be used to our advantage, in a sort of data science judo. We can do this in several ways, as I describe in the second part of the book. Here's one way.

I introduce so-called randomized controlled trials in chapter 2, and in chapter 9 I return to look at them from a different perspective. In a medical context, in the simplest such trial two treatments are compared by giving one treatment to one group of people and the other to another group. However, there is a risk. If the researchers know which treatment has been given to which people, this knowledge could influence the study. The researchers might be tempted to treat one of the groups more carefully than the other. For example, if the study aimed to compare an untested new treatment with the standard treatment, the researchers might monitor the former group for side effects more closely (perhaps subconsciously) or take more care in their measurement of possible outcomes. To overcome this potential bias, in such studies the treatment allocation is concealed from the researchers (*DD-Type 13: Intentionally Darkened Data*). The term *blinded* is used, to indicate that the data are dark.

Another familiar way in which dark data are used to advantage is in sample surveys. We might want to know the opinions of the people in a town, or of those who buy a company's products, and perhaps it is prohibitively expensive to ask them all. Certainly it is likely to be a time-consuming exercise, and opinions might change over its course. An alternative to asking everyone is to ask just some of them. The opinions of the others, the ones you don't ask, will be dark data. This might look like a high-risk strategy—it clearly resembles the TARN example. But it turns out that by using judicious methods of selecting which people you approach, you can get accurate and reliable answers—more quickly and cheaply than by attempting to approach everyone.

Yet a third way to use dark data to advantage is through so-called smoothing of data. As we shall see in chapter 9, this is equivalent to revealing unobserved and unobservable kinds of dark data (*DD-Type 14: Fabricated and Synthetic Data*), and it enables more accurate estimates and better predictions.

We shall explore other uses of dark data in chapter 9, where we will see that they often have exotic names. Some of them are widely applied in areas such as machine-learning and artificial intelligence.

All around Us

We've seen that dark data are ubiquitous. They can arise anywhere and everywhere, and one of their most dangerous aspects is that, by definition, we may not know that they are *not* there. It means we have to be constantly on the alert, asking ourselves, *what are we missing?*

Are we failing to notice large amounts of fraud because the police catch the inept criminals while the really good ones escape unnoticed? Bernie Madoff established his firm Bernard L. Madoff Investment Securities LLC in 1960 but wasn't arrested until 2008, and sentenced (to 150 years in prison) in 2009, when he was already 71—he almost got away with it.

Are we not noticing many potentially curable sick people simply because the more severe cases are obvious, but the less severe don't show so many symptoms?

Are the social networks established by modern social media dangerous simply because they reflect what we already know and believe, not challenging us because they don't show us facts or events outside our comfort zone?

Perhaps worse still, the descriptions people choose to post on social media may give us a false impression of how wonderful everyone else's life is, casting us into depression because in contrast our lives have so many obstacles.

We tend to think of data as numerical. But data don't have to be just numbers. And that means that dark data also don't have to be numerical. The following is an example in which the crucial missing information is a single letter.

The Arctic expeditions of 1852, 1857, and 1875 were stocked with a supply of Allsopp's Arctic Ale, an ale with an especially low freezing point prepared by brewer Samuel Allsopp. Alfred Barnard sampled the beer in 1889, describing it as "of a nice brown colour, and of a vinous, and at the same time, nutty flavor, and as sound as on the day it was brewed. . . . Owing to the large amount of unfermented extract still remaining in it, it must be considered as an extremely valuable and nourishing food."[10] Just the sort of thing you need to sustain you on Arctic expeditions.

In 2007 a bottle of the 1852 batch came up for sale on eBay, with a reserve price of $299. Or at least that was the aim. In fact the vendor, who had had the bottle for 50 years, misspelled the beer's name—he missed one of the *p*'s in Allsopp. As a consequence, the item didn't show up in the searches carried out by most vintage beer enthusiasts, so that there were just two bids. The winning bid, for $304, was from 25-year-old Daniel P. Woodul. Aiming to appraise the value of the bottle, Woodul immediately relisted it on eBay, but this time with the correct spelling. This time there were 157 bids, with the winning one being for $503,300.

That missing *p* clearly mattered, to the tune of some half a million dollars.* This shows that missing information can have significant consequences. In fact, as we shall see, a mere half-million-dollar loss is nothing compared with the losses that other missing data situations have led to. Indeed, missing data can wreck lives, destroy companies, and (as with the *Challenger* disaster) can even lead to death. In short, missing data matter.

In the case of Allsopp's Arctic Ale, a little care would have avoided the problem. But while carelessness is certainly a common cause of dark data, there are many others. The painful fact

*In fact it turned out that the winning bid was a practical joke, and the bidder had no intention of paying. But Woodul is nevertheless doubtless still sitting on a tidy profit: a private collector from Scotland recently auctioned a bottle from the 1875 expedition for £3,300 (~$4,300).

is that data can be dark for a tremendously wide variety of reasons, as we shall see in this book.

It is tempting to regard dark data as simply synonymous with data which could have been observed but which for some reason were not. That is certainly the most obvious kind of dark data. The missing salary levels in a survey in which some people refused to divulge how much they were paid are certainly dark data, but so also are the salary levels for those who do not work and hence do not have a salary level to divulge. Measurement error obscures true values, data summaries (such as averages) hide the details, and incorrect definitions misrepresent what you want to know. More generally still, any unknown characteristic of a population can be thought of as dark data (statisticians often refer to such characteristics as *parameters*).

Since the number of possible causes of dark data is essentially unlimited, knowing what *sort* of thing to keep an eye open for can be immensely useful in helping avoid mistakes and missteps. And that is the function of the *DD-Types* described in this book. These are not basic causes (like failure to include the final outcome for patients who have been in a study for only a short time) but provide a more general taxonomy (like the distinction between data we know are missing and data we don't know are missing). An awareness of these *DD-Types* can help in protecting against mistakes, errors, and disasters arising from ignorance about what you do not know. The *DD-Types,* which are introduced in this book, and which are summarized in chapter 10, are as follows:

DD-Type 1: Data We Know Are Missing
DD-Type 2: Data We Don't Know Are Missing
DD-Type 3: Choosing Just Some Cases
DD-Type 4: Self-Selection

DD-Type 5: Missing What Matters
DD-Type 6: Data Which Might Have Been
DD-Type 7: Changes with Time
DD-Type 8: Definitions of Data
DD-Type 9: Summaries of Data
DD-Type 10: Measurement Error and Uncertainty
DD-Type 11: Feedback and Gaming
DD-Type 12: Information Asymmetry
DD-Type 13: Intentionally Darkened Data
DD-Type 14: Fabricated and Synthetic Data
DD-Type 15: Extrapolating beyond Your Data

Chapter 2

DISCOVERING DARK DATA

What We Collect and What We Don't

Dark Data on All Sides

Data do not exist de novo. They have not been there since the beginning of time, sitting around just waiting for someone to come along and analyze them. Rather, someone has to collect them in the first place. And—as you doubtless expected—different ways of collecting data can lead to different kinds of dark data.

This chapter looks at the three fundamental ways that data sets are created, along with the dark data challenges associated with each method. The next chapter then explores some further dark data complications that can apply in many situations.

The three fundamental strategies for creating data sets are as follows:

1. Collect data on *everyone* or *everything* for the objects you are interested in.

   For human populations, this is what censuses strive to do. Likewise, stock-taking exercises aim to determine the details of everything in the warehouse or wherever. The annual stock take at London Zoo lasts about a week, revealing that (in 2018) there were 19,289 animals from species ranging from Philippine crocodiles to squirrel monkeys to Humboldt penguins and Bactrian camels (ants, bees, and other social insects are counted in colonies). In chapter 1 we noted that details of *all* the purchases in a supermarket are collected. The same applies to tax details and credit card transactions and company employees. Details of all sporting records, books on a library's shelves, prices in a shop, and so on, can be recorded. In all these examples the details of every one of the objects, people, or whatever are compiled together to form the data set.

2. Collect data on just *some* of the items in population.

   An alternative to a complete human population census is to collect data on just a sample of people. Survey sampling is very important in this sort of situation, and we shall look at it and its dark data issues in detail. Less formally, sometimes data are collected on what is readily available. To get a feel for how shoppers behave you might observe those who happen to come in today, or to see roughly how long it takes you to get to work each day you might record the daily times over a month. In other situations it is impracticable to measure everything: to see how food prices are changing over the course of time you cannot collect details of every food purchase made by everyone, and to determine the average weight of a grain of sand you cannot weigh each of them. In yet other contexts, as we saw in chapter 1, the very notion of measuring "everything" may be meaningless. You cannot record the results of all possible measurements of your height—only of those measurements you take.

Some years ago, before the era of readily available large data sets, my colleagues and I compiled a collection of 510 small real data sets that teachers of statistics could use to illustrate statistical ideas and methods (published as *A Handbook of Small Data Sets*[1]). Very few of those data sets describe complete populations. They include such things as the results of 20,000 dice throws, durations of pregnancies, corneal thickness in eyes, and nerve impulse durations.

3. Changing the conditions.

   The first two strategies for data collection are called "observational." You simply observe the values that the objects or people happen to have. In particular, you don't change the conditions under which you observe them, but measure them as they are. You don't give people a medicine and see how they respond, or ask them to perform a particular task and see how long it takes. You don't change the fertilizer to see if it makes crops grow more abundantly, or change the water temperature when you make your tea to see if it affects the taste. When you *do* change the circumstances under which you collect the data—when you *intervene*—the data are said to be "experimental." Experimental data are particularly important, because they can give you information about the counterfactuals mentioned in chapter 1.

Although these three modes of data collection have many dark data problems in common, they also lead to different kinds of problems. We begin by looking at comprehensive data sets.

Data Exhaust, Selection, and Self-Selection

The computer has revolutionized all aspects of our lives. Some of these ways are obvious, like the word-processing software I am using to write this book, or the travel booking system I use when I buy an air ticket. Others are concealed, like the computers controlling the brakes and engine of a car, or those inside an elaborate printer or photocopying machine.

But whether the role of the computer is obvious or not, in all cases the machine takes in data—measurements, signals, commands, or other kinds of data—and processes them to make a decision or carry out some operation. Then, once the operation has been completed, the processing could stop. But often it does not. Often those data are stored, sent to a database, and retained. They are spin-off data, data exhaust, which can be examined later to gain understanding, improve systems, and decipher what happened if things went wrong. Black box recorders on aircraft are classic examples of this sort of system.

When data like this describe humans, they are often called *administrative data*.[2] The particular strength of administrative data is that they actually tell you *what people do*, not (as can be the case with surveys, for example) *what people say they do*. They tell you what people bought, where they bought it, what they ate, what web searches they made, and so on. Administrative data, it is claimed, get you nearer to social reality than exercises involving asking people what they did or how they behave. This has led to the accumulation of giant databases describing our behavior by governments, corporations, and other organizations. There is no doubt that these databases represent a great resource, a veritable gold mine of potential value enabling all sorts of insights to be gained into human behavior. From those insights we can improve decision-making, enhance corporate efficiency, and devise better public policy—provided, of course, that those insights are accurate and have not been contaminated by the impact of dark data. Moreover, there are privacy risks which arise when data we would like to keep dark become known to others. We'll return to issues of privacy at the end of the section, but let's look first at unsuspected dark data.

One obvious high-level gap is that administrative data do indeed tell you what people actually do—which is useful, unless you actually want to explore what people think and feel. Discovering that a population of people in a particular corporation are unhappy with the way things are going might be just as important as noting how they behave under the constraints and imperatives of the corporation's daily activities with their boss looking over their shoulder. To discover how they feel we would have to actively elicit data from them—perhaps in a survey, for example. Different kinds of data collection strategies are suited to answering different kinds of questions—and have different kinds of dark data challenges.

My own first serious exposure to dark data was in the area of consumer banking: the world of credit cards, debit cards, personal loans, auto finance, mortgages, and so on. Credit card transaction data involve giant data sets, with millions of customers making billions of plastic card transactions each year. Around 35 billion Visa card transactions were made between June 2014 and June 2015, for instance.[3] Whenever a purchase is made with a credit card, details of the amount spent, its currency, the vendor, the date and time of the transaction, and many other items of information are recorded (in fact, 70–80 items of information). Much of this information has to be collected so that the transaction can be made and the appropriate account charged; it's a necessary part of the operation, so that omitting these details is unlikely or even impossible. For example, the transaction could not take place without knowledge of how much to charge or who to charge it to. But other items of data might not be critical to the operation, so it is possible they might not be recorded. For example, omitting invoice numbers, detailed product codes, and unit prices would not interfere with the operation. Clearly this is an example of our first dark data type: *DD-Type 1: Data We Know Are Missing.*

Worse still, at least from a dark data perspective, while some customers will use a credit card for their purchases, others might use cash. This would mean that, as a record of *all* purchases and transactions, the credit card database would have unseen swaths of dark data, arising because of *DD-Type 4: Self-Selection.* Moreover, there are multiple credit card operators. The data from one operator may not be representative of the entire population of credit card holders, and certainly not of the entire population altogether. So, while holding great promise, administrative data might well have dark data shortcomings which are not obvious at first glance.

The particular problem I was presented with was a request to construct a "scorecard"—a statistical model for predicting whether an applicant was likely to default with repayments which could be used to guide the decision about whether the bank should give him or her a loan. I was supplied with a large data set giving the application-form details of previous customers, along with the outcome indicating whether or not those previous customers had actually defaulted.

In essence the exercise was straightforward. I needed to find what patterns of characteristics distinguished those customers who had defaulted from those who had not. Then future applicants could be categorized by determining whether they were more similar to the defaulters or the nondefaulters.

The trouble was that the bank wanted to make predictions about *all* future applicants. The data given to me were surely unlike the population of future applicants, because my data had already gone through a selection process. Presumably the previous customers had been given a loan because they were thought to be good risks according to some earlier mechanism—either a previous statistical model or perhaps a bank manager's subjective opinion. Those previously thought to be bad risks would not have been given a loan, so I knew nothing about whether they would actually have defaulted. Indeed, I had no idea how many applicants had previously been declined and not made it into my data set. In short, the data given to me were a distorted sample, subject to an unknown extent of selection or selectivity bias, and any statistical model built on this distorted data set could be very misleading when applied to the overall population of potential future applicants.

In fact, the problem was even worse than that. It actually had multiple layers of dark data. Consider the following:

Who actually applied? In the past the bank might have mailed potential customers asking if they would like a loan. Some would have replied they did want a loan, and others would not have replied. The data would include only those who had felt motivated to reply to the initial mailshot, and this might depend on how it was worded, how much was offered, the interest rate, and a host of other factors about which I knew nothing. The ones who had not replied would represent dark data.

Who received an offer? Those who replied would have been evaluated, and some of those would have been offered a loan, while others would not. But since I didn't know on what basis this offer had been made, I was presented with more dark data.

Who took up the offer? In addition to the preceding two selection processes, of those who had been offered a loan some would have taken it up, while others wouldn't have—introducing yet another layer of dark data.

Adding all these layers together made it very unclear how the data I was given related to the problem to be solved, which was to build a model to evaluate new applications. The multiple layers of dark data could mean that the sample I had, with all the known good/bad actual outcomes, was completely different from the population to which the bank would like to apply the model. Ignoring the dark data could be disastrous. (The bank still exists, so I suppose my model was not that bad!)

Administrative data are ubiquitous—just think of all the databases storing information about you relating to education, work, health, hobbies, purchases, financial transactions, mortgages, insurance, travel, web searches, social media, and so on. Up until very recently, in most of these cases your data were stored automatically, without your knowing about it and having a say in it. The European Union's General Data Protection Regulation (GDPR) has changed that—as you doubtless realize because of all the invitations to check boxes saying you understand and give permission for personal data about you to be recorded by websites. But occasionally you can have a say in other ways as well. (The protection of data of U.S. residents is regulated by both federal and state laws, varying by sector.)

In 2013 the UK National Health Service (NHS) launched a scheme whereby medical data would be copied from family doctor records each month and merged with hospital records in the national Health and Social Care Information Centre (HSCIC).

The potential value of such combined data sets is immense. By pulling together data describing the medical conditions and treatments of millions of people it becomes possible to mine the data to understand not only the medical conditions themselves, including preventing illness and monitoring diseases and the effectiveness of treatments, but also how effective healthcare delivery is across the system and where it needs to be improved. Confidentiality would be preserved by using a system of "pseudonymization," in which names, NHS number, and other identifiers were replaced by a code, and the codes were stored in a file in a different place from the actual data.

Unfortunately, the system—and its potential medical and health benefits—was poorly presented to the public. Some people were anxious that their data might be sold to commercial third parties (like pharmaceutical and insurance companies) which would seek to use the data for a profit. Others were worried about data loss and hacking, and that their data might be *reidentified*, linked back to them, with implications for personal privacy about confidential medical matters. As a result there was a public backlash, which was reinforced by some parts of the media. This was true even though the system allowed people to opt out, that is, allowed them to say they did not want their data to be transferred.

In February 2014 the program was put on hold. Eventually, after a number of false starts, an extensive review was published in July 2016, recommending an eight-point consent model for patient data sharing. One of its features was that it gave people the power not to allow their data to be used for purposes beyond medical care, such as for research.

If you have been following the discussion so far, you will probably have spotted the danger. Allowing people to opt out means that the databases are not comprehensive. They contain information on only some of the patient population. Worse, since people themselves choose whether or not to have their data included (*DD-Type 4: Self-Selection*), it is entirely possible that the databases show a distorted picture of the population.

In 2009, Michelle Kho and colleagues from McMaster University looked at these sorts of issues.[4] They carried out a meta-analysis of studies exploring whether requiring informed consent for data to be used in this way influences who will agree to take part. Looking at studies using medical records, they compared consenters and nonconsenters in terms of age, sex, race, education, income, and health status, and they found that the two groups did differ. But alarmingly they also found that "there was a lack of consistency in the direction and the magnitude of effect." That means that not only did consenters and nonconsenters differ, but they differed in unpredictable ways, making it very difficult to adjust for the difference.

Opting out—allowing people the right not to be included in a database—at least requires them to make an effort to be excluded. In such circumstances many people are lazy, going with the default, and so are included in the database. A potentially more serious alternative is requiring people to opt in—to make the effort if they want to be *in*cluded in a database. Now, intrinsic laziness can make things worse: requiring people to make an active effort is a sure way of reducing the response.

That medical record example involved very explicit administrative data. Sometimes things are not so obvious. Take the example of abandoned phone calls to the emergency services.

An abandoned call is one in which someone calls the emergency number but hangs up or terminates the call before being connected to the operator. In September 2017, the BBC website reported that over the year starting June 2016 the number of abandoned emergency calls to UK police control rooms doubled, from 8,000 to 16,300.[5] There are various theories about why this might be happening, but one is that overstretched police forces are taking too long to answer. Another is that mobile phones are automatically generating such calls, perhaps as the buttons are accidentally pressed in a pocket or handbag.

If that last theory were the sole cause, we might expect the problem not to arise, or at least be less serious, in the United States, where the emergency call number 911 uses two different digits (it's 999 in the UK). But the rate of such calls is rising in America also. Records over three months from the Lincoln Emergency Communications Center illustrate the sort of change, with the percentage of abandoned incoming calls increasing from 0.92 percent to 3.47 percent from April through June 2013.

Abandoned calls are a clear case of *DD-Type 1: Data We Know Are Missing*. In contrast, a wonderful example of *DD-Type 2: Data We Don't Know Are Missing* was given by Mike Johnston in his column *The Online Photographer*.[6] He wrote: "I have to chuckle whenever I read yet another description of American frontier log cabins as having been well crafted or sturdily or beautifully built. The much more likely truth is that 99.9 percent of frontier log cabins were horribly built—it's just that all of those fell down. The few that have survived intact were the ones that were well made. That doesn't mean all of them were." Since there is no record of all the many log cabins which have collapsed and decayed, these are dark data.

DD-Type 2: Data We Don't Know Are Missing is particularly deceptive because we will generally have no reason to suspect it. Suppose, for example, we read, as I did in the *Times* (London) of 29 December 2017, that "the number of sexual assaults allegedly carried out by taxi drivers on passengers has risen by a fifth in three years, according to police figures." The immediate and superficial explanation is that more such offenses are being committed. But there is an alternative explanation, arising from dark data. This is simply that while the rate of commission of such offenses is remaining constant the rate of *reporting* of the offenses is increasing. Hitherto concealed dark data may be becoming visible as a result of changing social mores and societal norms. There's a general moral there: if you see a sudden step change in a time series of values, it could be because the underlying reality has changed, but it could also be because the data collection procedure has changed. This is a manifestation of *DD-Type 7: Changes with Time.*

A more elaborate example of *DD-Type 2: Data We Don't Know Are Missing* and *DD-Type 7: Changes with Time* working in tandem is illustrated by the performance of investment funds. The population of such funds is dynamic: new funds are set up, and old ones die. And, unsurprisingly, it is generally the underperforming funds which die, leaving just the ones which do well. Superficially, if we do not somehow take those that drop out into account, on average such funds will appear to do well.

Although individual funds which have dropped out because they performed badly will be excluded from an index showing overall or average performance, it might be possible to look back and obtain data on those funds. This would change them from *DD-Type 2: Data We Don't Know Are Missing* to *DD-Type 1: Data We Know Are Missing,* and it would then be possible to explore the impact of excluding them from the calculations. A 2006 study by Amy Barrett and Brent Brodeski showed that "purging of the weakest funds from the Morningstar database boosted apparent returns on average by 1.6 percent *per year* over the 10-year period [from 1995 to 2004]."[7] And in a study published in 2013, Todd Schlanger and Christopher Philips of Vanguard looked at the performance of funds including and excluding closed funds over 5-, 10-, and 15-year periods.[8] The differences were striking, with the performance of those excluding the closed funds over 15 years being almost double that when they were included. This study also revealed the magnitude of dark data in this context: only 54 percent of funds lasted the full 15-year period.

The phenomenon also affects more familiar financial indexes such as the Dow Jones Industrial Average and the S&P 500. Companies which perform poorly drop out of these indexes, so that only those which do relatively well contribute to the final performance value. This is fine if you happened to have invested in those companies which went on to do well, but not so fine otherwise. And since it's very difficult (some would say impossible) to tell which companies are going to go on to do well and which aren't, the index performance is deceptive.

Having cautioned about so-called survivor bias in financial indexes, it is worth noting that things can be more complicated. Taking hedge funds as an example, certainly, poorly performing funds are likely to close and not be included in the data, but so also are funds at the opposite end of the spectrum: exceptionally strongly performing funds are likely to close to new investors. Likewise, strongly performing companies can split and so drop out of a share index. Dark data can work in mysterious ways.

Additionally, for reasons we shall explore in chapter 3, there is a good chance funds which have performed exceptionally well in the past will nosedive in the future owing to the phenomenon of "regression to the mean." This means purchasers of funds need to look very carefully at how past performance is evaluated. As in other walks of life, investors need to ask themselves if the truth is being disguised by invisible dark data.

Survivor bias is always a potential problem for things which change over time. In the world of startups we tend to hear more about the successes than the failures—even though the majority of such companies fail. Some researchers put this failure rate as low as 50 percent, while others put it as high as 99 percent. Of course, it partly depends on the time period you are considering (one year, 50 years?) and how you define "failure." Take the social networking site Bebo, for example. Launched in 2005, at one stage Bebo was the most popular social networking site in the UK, with nearly 11 million users. In 2008 it was bought by AOL for $850 million. So, over a three-year horizon Bebo was hugely successful. But then the number of users started to fall, partly as they shifted to Facebook, and in 2010 AOL sold Bebo to Criterion Capital Partners. A computer glitch damaged its reputation, and in 2013 Bebo filed for Chapter 11 bankruptcy protection. Later in 2013 the original founders, Michael and Xochi Birch, bought the company back for $1 million. So is this a success or a failure? And what about Lehman Brothers? This firm was founded in 1850 and became the fourth largest investment bank in the United States—until it filed for bankruptcy in 2008, that is. Like Bebo, the company came to a sticky end, albeit over a longer time interval. But was it a success or a failure?

In the startup world people would naturally *like to* hear about the success stories more than about the failure stories, simply because they are trying to emulate the successes and not the failures. But this situation reveals another kind of dark data. What entrepreneurs should be looking for are characteristics which *distinguish between* successes and failures, not simply characteristics which happen to have been associated with successes. Characteristics of the latter kind might also be associated with the failures. Moreover, even if the characteristics are associated with successes more than failures, there is no guarantee they are causal.

The wonderful comic website *xkcd* has a cartoon about survivor bias.[9] The character is advising us never to stop buying lottery tickets, describing how he lost and lost time after time but kept buying tickets, even taking extra jobs to earn money to buy more tickets. And he eventually succeeded (if "succeeded" is the right word). What we don't see are the gamblers who poured fortunes into lottery tickets but died without winning.

In general, administrative data have immense potential to do good, provided we appreciate the dark data risks. But there is a complementary aspect which might be less positive that is leading to increasing concern.

From our individual perspective the data exhaust retained in an administrative data database is a *data shadow*. It consists of the traces we leave from sending emails or texts, tweeting, posting a comment on YouTube, swiping credit cards, using travel cards, making phone calls, updating a social media app, logging onto a computer or iPad, taking cash from an ATM, driving past a car license plate recognition camera, and so on endlessly, in often unsuspected ways. While such data can indeed be aggregated to benefit society, they also inevitably reveal a huge amount about each of us as individuals, our likes and dislikes, and our habits and behaviors. The data relating to us as individuals can be used to our benefit—guiding us toward products or events which might interest us, facilitating travel, and generally smoothing life out for us. But they can also be used to manipulate behavior. Authoritarian regimes can exert considerable control over us if they know detailed patterns of our lives. In a way this is inevitable: the downside of giving out information so that we can be assisted is that . . . we give out information.

Because of increasing concern about data shadows, services exist which will minimize our shadow. Or, from the perspective of this book, services exist to switch off the light on data, rendering them dark. Basic steps include deactivating all social media accounts (Facebook, Twitter, etc.), deleting old email accounts, deleting search results, using false information for accounts we cannot delete (e.g., false birth dates or middle initials), unsubscribing from lists and alerts, and so on. Of course, the complementary side of the protection created by hiding data is that the potential benefits will be adversely impacted. We can determine whom to give tax credits to only if we have details of people's income and tax payments.

From the Few to the Many

Collecting data on *all* the people or objects we are interested in—like the administrative data generated during supermarket purchases—is a powerful way of acquiring information that can improve our understanding, enabling us to make better decisions. But such data are not always adequate for shedding light on any particular question we might want to answer. The obvious example is simply the situation in which no relevant administrative data sets are being automatically compiled. One solution might be to use the most closely related data set we can find, but that has its own risks. A second solution would be to deliberately carry out a separate exercise to collect data on the entire population aimed specifically at answering the question. That is, to carry out a census. Unfortunately, censuses are typically expensive and slow processes; and there's no point in spending a vast amount of money to get a perfect answer long after it has ceased to be of use.

A third possible strategy is to use a survey.

Surveys are one of the main tools used for understanding modern society. Their particular strength is that they enable us to find out what's going on in a population or group without asking everyone in that population or group. They are based on a very powerful statistical phenomenon called *the law of large numbers,* which says that the average value of a (randomly drawn) sample taken from a population has a high probability of being very close to the real average value of the population if the sample size is large enough.

Suppose we want to find the average age of people in a country. This matters because it is a basic piece of information in determining whether there are (and will be as the population ages) sufficient younger, wage-earning and tax-paying workers to support the older retired population. Its importance and potential impact is illustrated by contrasting the West African country of Niger, where 40 percent of the population is aged less than 15, with Japan, where just 13 percent is.

Suppose we don't have birth records, that we cannot afford to conduct a census to find out everyone's age, and that we are uneasy about the dark data associated with administrative data in the many other databases which requested date of birth on signing up for some service. Surveys allow us to be confident of getting a sufficiently accurate estimate by asking only *some* of the people their age. You might be immediately suspicious of this: there's the obvious dark data risk of not knowing the age of all those people we don't ask. But the law of large numbers tells us that it is possible, provided the sample is properly drawn. Moreover, the mathematics behind this law also tell us that the sample need not be very large—a thousand might be large enough. Obtaining data from a thousand people is a very different matter from obtaining it from the possibly many millions of people in the entire population.

The throwaway comments describing the sample as "randomly drawn" and "properly drawn" are critical. If we took our sample solely from people in nightclubs or in retirement homes, we would probably not get a very accurate estimate of the overall population average age. We need to be as confident as we can that the sample is properly representative of the population we are studying. The best way to achieve that is to begin by listing all the members of the population we are interested in—such a list is called a *sampling frame*—and to randomly choose a sample of people from that list and ask them their age. Such comprehensive lists are often available from administrative data. For example, electoral rolls are often used for such lists, as are earlier censuses.

At first the notion of *randomly* choosing whom to ask their age might seem bizarre. Surely that means each time we undertook such an exercise we could get a different result. That's true, but while this approach will not *guarantee* that there are no unfortunate dark data distortions in the sample (e.g., that it does not have a higher proportion of young people than is in the population), the *probability* of such distortions can be kept to a controlled level. It means we are able to make statements like "for almost all (e.g., 95 percent) of the samples we could have drawn, the sample average will be within two years of the population average age." And by increasing the size of our sample, we can increase our confidence from 95 percent to 99 percent or indeed whatever level we think is appropriate, and likewise we can make the interval two years or one year or again whatever we want. And in case you are concerned about the lack of absolute certainty in the conclusions drawn by such a process, perhaps I should remind you that *nothing* in life is absolutely certain (with the possible exceptions of death and taxes).

One of the interesting aspects of the law of large numbers is that the accuracy of the estimate doesn't materially depend on how large a fraction of the population is included in the sample, at least if the population is large and the sample a relatively small fraction of it. Instead, it depends simply on how large the sample is. Other things being equal, a sample of a thousand drawn from a population of a million will usually give much the same accuracy as a sample of a thousand drawn from a population of a billion. This is true even though the sample is one in a thousand of the first population and one in a million of the second.

Unfortunately, this sample survey strategy is not a magic wand. Like (almost?) everything else in life, there is a downside, namely, that surveys generally rely on voluntary participation. That means that people might answer some of the questions but not others, or they might even decline to take part at all. We have entered the realm of dark data (*DD-Type 4: Self-Selection*).

An example of such "nonresponse" is given in Table 1, which shows a sample of data with some values missing. These are indicated by a "?" symbol (often "NA" is used, for "not available"), and these are highlighted so they are easier to see. The data are a list of 10 records from the marketing data given on the Knowledge Extraction Based on Evolutionary Learning website.[10] They were collected from questionnaires given to people in a shopping mall in the San Francisco Bay area, and the aim of the study was to construct a model to predict income from the other variables. The variables are given as *A*=Sex, *B*=MaritalStatus, *C*=Age, *D*=Education, *E*=Occupation, *F*=YearsInSF, *G*=DualIncome, *H*=HouseholdMembers, *I*=Under18, *J*=HouseholdStatus, *K*=TypeOfHome, *L*=EthnicClass, *M*=Language, and the variable to be predicted was that in the final column, *N*=Income (the website gives more details of the meaning and range of each of the variables; I have introduced the letter labels for convenience). In all, the complete data set contains 8,993 rows like those shown in the table, but 2,117 of them have missing values—as do three of the rows of the extract shown. One of these three has two values missing. The values missing in the table are clearly *DD-Type 1: Data We Know Are Missing,* as we can see that there should be something there.

TABLE 1. Extract from marketing data

	A	*B*	*C*	*D*	*E*	*F*	*G*	*H*	*I*	*J*	*K*	*L*	*M*	*N*
Rec 1	2	1	3	6	6	2	2	4	2	1	1	7	1	8
Rec 2	2	1	5	3	5	5	3	4	0	2	1	7	1	7
Rec 3	1	4	6	3	?	5	1	1	0	1	1	7	?	4
Rec 4	2	1	5	4	1	5	2	2	2	1	1	7	1	7
Rec 5	2	3	3	3	2	2	1	2	1	2	3	7	1	1
Rec 6	2	1	5	5	1	5	2	2	0	1	1	7	?	9
Rec 7	2	1	5	3	5	1	3	2	0	2	3	7	1	8
Rec 8	1	5	1	2	9	?	1	4	2	3	1	7	1	9
Rec 9	1	3	4	2	3	4	1	2	0	2	3	7	1	2
Rec 10	2	1	4	4	2	5	3	5	3	1	1	5	2	9

Source: Website Knowledge Extraction Based on Evolutionary Learning, http://www.keel.es/.

The column labeled *M*, which is Language, is the answer to the question "What language is spoken most often in your home?" with the three possible response categories being 1=English, 2=Spanish, and 3=Other. Since only one of these categories can be true for each household and since these categories include all possible answers (e.g., if German was the language spoken most often, the response would be 3=Other), we know that there really is an answer—1, 2, or 3—for this question for each row. It's simply that, for some reason, two people in our sample have not given it.

Sometimes, however, the records are incomplete because *there simply are no values to be put there*; the answer really does not exist. For example, a form which requested the age of the respondent's spouse would necessarily have a blank if the respondent was not married. That poses the interesting question of how we should treat that sort of missing value. The situation is clearly different from one in which someone is married but simply has not filled in that field in the form. But does this difference matter?

If the two types of nonresponse are treated in the same way, will it lead to mistaken conclusions?

A record which has gaps shows immediately that something is missing (*DD-Type 1: Data We Know Are Missing*). But what about people who refuse to answer any questions at all? The result is *DD-Type 4: Self-Selection*. People might be too busy, feel it too much of an invasion of privacy, or simply be uncontactable (e.g., they are out of town when the survey is conducted). These are also known unknowns in that we know who they are—they are on the list of people we intended to contact—and that they *could have* given answers if they had been willing and able. But since we don't have any of their responses, we really are in the dark.

One of the most famous examples of this sort of problem occurred in the 1936 U.S. presidential election. On the basis of polls, the popular magazine *Literary Digest* had previously very successfully predicted winners of elections, and in 1936 it predicted that the Republican candidate, Alfred M. Landon, would win by a 3 to 2 majority. However, Franklin D. Roosevelt, the Democratic candidate, won in a landslide, garnering 523 of the 531 votes of the Electoral College, 62 percent of the popular vote, and 46 out of 48 states.

Descriptions of this election outcome and the mistaken *Literary Digest* prediction often attribute the failure to dark data arising from the design of the survey. In particular (though there are slightly different stories, they all have the same core idea), phone directories were used as the sampling frame from which potential respondents were chosen. At the time, however, phones were something of a luxury and were more likely to be possessed by more affluent people, who were more likely to vote Republican. Thus, the sample would have overestimated the proportion of people who would vote Republican.

Much like the example of trying to estimate the average age of a country's population based on samples drawn from nightclubs or retirement homes, this explanation for the election forecasting failure suggests that the cause of the dark data lay not in people's refusal to answer questions but rather in a faulty selection of the initial list of people who were to be interviewed.

However, a detailed analysis by statistician Maurice Bryson shows that this simplistic analysis is wrong.[11] On the one hand it underestimates the lengths to which the *Literary Digest* pollsters went to ensure representativeness of the sample of potential voters they approached. They were very much aware of the factors which could lead to sample distortion. On the other hand, although only about 40 percent of householders at the time had phones, these would also be the ones most likely to vote. The latter factor meant that although there might be a mass of dark data from the population as a whole, it would be less serious when thought of as a proportion of the *voting population*, which is of course what counts in an election. Incidentally, this can have a major impact: in the 2016 UK referendum about whether to leave the European Union, 43 percent of those who claimed "no interest in politics" voted, as opposed to just 30 percent voting in the 2015 UK General Election. A poll assertion that someone *will* vote for X needs to be translated into an action that they *do* vote for X.

Concerning the Landon/Roosevelt election poll, it seems that although the phone explanation is the one which is widely believed and propagated, it is wrong.

What, then, explains the failure of the poll?

The answer still lies in dark data but in a much more familiar and straightforward form—one which has become very important with the advent of web-based surveys. This is that although the pollsters mailed out 10 million ballots, only about a quarter of those who received them, some 2.3 million people, bothered to reply. Over three-quarters of those approached simply ignored the ballot—their views became dark data. The implications are obvious. If (as seems to have been the case) the Republican voters were more interested in the election than the pro-Roosevelt voters, they would have been more likely to respond to the poll. That meant that the poll would give the impression of a majority in favor of Landon—the distorted view that the *Literary Digest* saw. This self-selection distortion vanished when the actual election was held.

So this polling disaster was not a consequence of dark data due to the poll designers' errors in drawing the sample but of dark data arising from the different probability that Republican and Democrat voters would respond to the survey. It was a consequence of *voluntary response* (*DD-Type 4: Self-Selection*).

Since there was a properly defined sampling frame—that is, the pollsters knew precisely who was eligible to vote—with a sophisticated enough analysis it might have been possible to adjust for it, as described in chapter 9. But in cases where there is no properly defined sampling frame, such adjustment will be much more difficult or simply impossible. If there's no sampling frame, then we have left the domain of *DD-Type 1: Data We Know Are Missing* and moved into that of *DD-Type 2: Data We Don't Know Are Missing*. Web surveys, which we discuss later, are particularly vulnerable to this scenario.

In general, situations in which some people do not respond at all, as in the Landon/Roosevelt example, are tough to handle. It could very easily be that those who refuse to respond differ in important ways from those who do choose to respond: the simple fact that they have chosen not to engage with the survey shows they are different in some way. Perhaps people with a particular interest in the topic of the survey might be more inclined to respond, as in the Landon/Roosevelt case. Or perhaps those who are more likely to respond might be better informed about the topic. One Dutch study of housing demand found that people were more likely to respond to a survey if their needs were greater, so that the survey was in danger of giving a very misleading overall impression.[12] In victimization surveys, so-called series events, such as spousal abuse, which have no clear-cut start and end point might not be picked up by surveys which focus on individual incidents. In other contexts, people might not wish to engage if the perceived time burden is too great. Overall, self-selection is a particularly pernicious generator of dark data in surveys and in other situations.

Pre-election polls are useful, but since surveys are used very widely to collect information in government and commerce, inadequate response rates can have widespread implications. And response rates are falling worldwide. An illustration is provided by the UK's Labour Force Survey.[13] Figure 2 shows the fraction of those approached who agreed to be interviewed in each quarter from March–May 2003 to July–September 2017. The proportion fell from over 65 percent to below 45 percent over this period. The straight-line downward trend does not lead to a great deal of confidence in the conclusions which will be drawn from this survey in the future, unless action is taken to improve the response rate.

The phenomenon is not restricted to the Labour Force Survey in Britain but affects surveys of all kinds worldwide. The U.S. Survey of Consumer Attitudes is a telephone survey of attitudes to spending and the economy. Its participation rate declined from 79 percent in 1979 to 60 percent in 1996 to 48 percent in 2003. Many further examples are given in a 2013 report from the U.S. National Academies edited by Roger Tourangeau and Thomas Plewes which says: "Response rates to government and

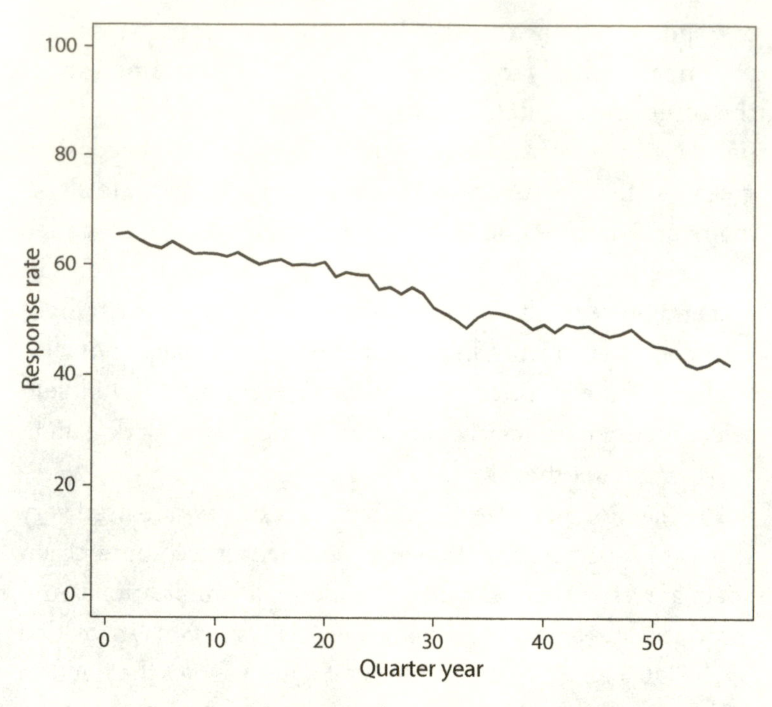

FIGURE 2. Proportion who agreed to be interviewed in the UK Labour Force Survey each quarter from March–May 2003 to July–September 2017.

privately sponsored household surveys that provide rich data for social science research have been falling throughout the richer countries of the world."[14] Figure 3 shows the response rates from 1997 to 2011 for households within the U.S. National Health Interview Survey. Although perhaps not as pronounced as in the British example, the downward trend is still apparent.

The decline is also evident in medical epidemiological studies. The Behavioral Risk Factor Surveillance Survey is used nationwide in the United States to study risk factors, health screening, and healthcare access. The median participation rate in this survey dropped from 71 percent in 1993 to 51 percent in 2005.

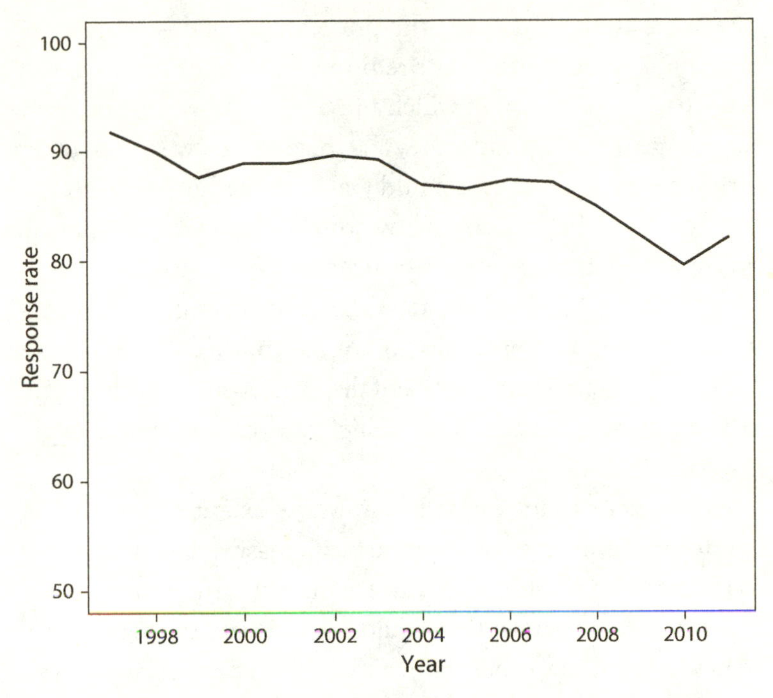

FIGURE 3. Response rates from 1997 to 2011 for households within the U.S. National Health Interview Survey.

The key question is, At what point does the response rate become too low for a survey to be useful? At what point does the proportion of data which are dark become too high for the results of the survey to be trusted as a summary of the overall population? Is a 90 percent response rate good enough to produce reliable results? If so, what about 80, 50, 20 percent? And how effective are methods (like those described in chapter 8) for adjusting the results to allow for the nonresponse?

Unfortunately, there is no general answer to these questions. It will depend on what the survey is about, on what questions are to be asked, and on how and why the data are missing. In some instances even a small proportion of missing records could mean the available data are unrepresentative of the overall population. A survey of the entire population about attitudes to sex-change operations in which some question was so offensive to transgender people that they all refused to answer while other groups were not affected could yield misleading results, even though it would have only a low nonresponse rate. In contrast, in other instances, even a large proportion of missing records might have only a small impact on the conclusions. In fact, it follows from our earlier discussion of the importance of sample size and random sampling that if the nonrespondents don't fall into any particular pattern, then a high rate of nonresponse might not matter at all.

In any case, as the examples show, the extent of dark data in the form of nonresponse seems to be increasing, and while those dark data might not have an adverse impact on any conclusions, in some circumstances they could have a very serious impact. If you were running a country or a major corporation, would you want to rely on "might not"?

One of the exciting possibilities of the internet is that it facilitates low-cost surveys, with the potential to reach vast audiences and hence huge sample sizes. But there are concomitant shortcomings. In particular, there is a lack of control over who responds to the questions. To a large extent, they will be self-selected, deciding for themselves whether to take part in a survey or to vanish into the opacity of dark data. This is obviously potentially disastrous to any kind of survey, since it means that the responses are likely to be related to whether someone answers the survey questions at all. (Recall the hypothetical magazine survey example from chapter 1 asking the question "do you reply to magazine surveys?") This is on top of distortion arising from who sees the relevant web page in the first place.

At the other extreme, people might even be able to respond to a web survey multiple times. Or even worse: while writing this book I came across one person who said that whenever he was presented with a survey on his mobile phone he handed it over to his five-year-old son to answer. Moreover, a fundamental problem is that not everyone has access to the internet—compare the role of telephones in the Landon/Roosevelt election. A 2013 Dutch paper published in the *International Journal of Internet Science* reported that "older households, non-Western immigrants, and one-person households are less likely to have Internet access."[15] As if all that wasn't problematic enough, access is likely to change over time as the population ages, and indeed as technology progresses.

Underlying nonresponse is the question of *why* people appear to be decreasingly willing to answer surveys. Tourangeau and Plewes looked into this question and found that the reasons for nonresponse hadn't changed much over time.[16] The main ones were that the potential respondent was simply not sufficiently interested to engage, was too busy, or that the interview would take up too much time. Other reasons included privacy concerns or not understanding the survey, and negative reactions included hanging-up, door-slamming, or hostile or threatening behavior. Clearly, being an interviewer can be a risky business! It has been suggested that people have simply become saturated with surveys—there are too many of them and people are exasperated at answering endless questions. The situation is aggravated by commercial promotion campaigns masquerading as surveys. The metalevel reason behind all this is self-selection; that is, the respondents are deciding for themselves whether to take part.

Survey nonresponse is not necessarily due to the respondent. It might be that an interviewer failed to make sufficient effort to contact individuals. Indeed, dishonest interviewers have even been known to make up data (*DD-Type 14: Fabricated and Synthetic Data*). The term "curbstoning" has been coined to describe this occurrence, reflecting (in caricature) the notion that instead of going to the trouble of asking people questions, a census enumerator might simply sit down on the curb and make up the numbers.[17] Be warned, however, that sophisticated statistical methods can often detect this deception, as they can other kinds of data fraud. Language barriers and simple data loss can also lead to missing values.

Sensitive survey questions—perhaps those related to sexual activity or financial or medical matters—can be particularly susceptible to missing and incomplete records. Consequently, some rather clever data collection methods have been developed which allow people to answer such questions while retaining their anonymity or in such a way that aggregate statistics can be produced without individual values being known. We shall look into these in chapter 9.

Experimental Data

We have looked at two of the modes of data collection described at the start of this chapter: situations in which "all" the data were recorded, and situations in which a sample of data was used. And in particular we have explored how different kinds of dark data can arise in each situation.

We now turn to the third mode of data collection, the experimental mode, in which the circumstances, treatments, or exposures that objects (or people) are subject to are varied in a carefully controlled way.

Suppose we wish to know which of two medical treatments is the more effective. Let's call the treatments A and B. The obvious place to start is by giving both treatments to a patient and seeing which works better.

That's useful if it is possible to give a patient both treatments. For example, in a study of the effectiveness of medications to alleviate hay fever symptoms, we could give the patient treatment A one year and treatment B the next, or the other way around (hoping that the pollen count was the same in the two years). But in many cases it's not possible to give a single patient both treatments. For example, in a study to see how effective treatments are at extending life expectancy, we can hardly go back and try the other treatment once the first has reached a conclusion and the patient has died.

The only alternative to giving a single patient both treatments is to give one patient treatment A and another patient treatment B. The trouble is, of course, that not all patients respond the same way to a given treatment: just because one patient did well under treatment A does not mean all patients will do well. Indeed, even worse, any particular patient might respond differently to the same treatment at different times.

So we now go further. First, we shift our attention from individual patients to the *average* responses of several patients receiving each treatment: how well *on average* do patients respond to A compared with how well they respond on average to B? And second, we try to ensure that the patients are allocated to the two treatments in such a way as to ensure that other factors cannot account for any differences between the results. For example, we would not want to give all the men one treatment and all the women the other because we would then not know if any observed differences in outcome were due to the treatments or to sex. Likewise, we wouldn't want to give all the sicker patients treatment A and the less sick patients treatment B.

So, we could balance the allocations. For example, we could give half of the men one treatment and the other half the other treatment, and the same for the women, and half of the sicker patients one treatment and the other half the other treatment. This would be possible if we wanted to control for just a limited number of factors. Thus, we could try to balance factors such as sex, age, and severity of illness, but we would soon find the task starts to become impossible: perhaps there is no seriously ill 25-year-old man with high blood pressure, body mass index of 26, a history of asthma, and who smokes, to balance against a seriously ill 25-year-old woman with exactly the same conditions. Worse still, there will doubtless be all sorts of factors we hadn't thought of.

To overcome this problem, people are *randomly* assigned to the two treatment groups—hence *randomized controlled trials* (RCTs). In this way we can be sure that the *chance* of imbalance of the kind we want to avoid is very small. You will recall the rationale behind drawing a random sample when we looked at surveys earlier. Exactly the same sort of principle is in play here, except that now we are allocating patients to treatments instead of simply selecting which people to interview.

Comparing just two groups in this way is a very simple experiment. Sometimes it's called an A/B trial, with the two groups labeled A and B, or a champion/challenger trial, in which a new method or treatment (the challenger) is compared with the standard method (the champion). Such designs are very widely used, in medicine, web experiments, and manufacturing, for example. The strength of such studies lies in the fact that we can compare the outcome with *what would have happened otherwise.* We have sidestepped the counterfactual dark data problem.

This way of controlling what data are dark has been around for a long time. One of the early modern influential randomized controlled trials was the 1948 study of the treatment of tuberculosis by streptomycin. Of this study, British health service researcher Sir Iain Chalmers said: "The detailed and exceptionally clear 1948 report of the British Medical Research Council's randomized trial of streptomycin for pulmonary tuberculosis is rightly regarded as a landmark in the history of clinical trials."[18]

As is often the case, however, it is possible to trace the ideas, or at least roots of them, throughout history. An earlier description of randomizing the group allocation includes the suggestion by the Flemish physician Jean-Baptiste Van Helmont, made in 1648, to decide the effectiveness of bleeding and purging patients: "Let us take out of the hospitals . . . 200 or 500 poor people, that have fevers, pleurisies. Let us divide them into halves, let us cast lots, that one halfe of them may fall to my share, and the other to yours; I will cure them without bloodletting and sensible evacuation; but you do, as ye know. . . . We shall see how many funerals both of us shall have."[19] You may be unfamiliar with this old-fashioned use of the word "lot," but you will certainly be familiar with it in the word "lottery."

So far so good. Unlike the other two data collection modes, which simply collected data on how people behaved ("observational data"), this new approach controls who receives what treatment. And that would be fine if all the patients stuck to the treatment protocols, taking the medication when they were supposed to and continuing until the end of the trial. But unfortunately, dark data often occur in such trials in the form of *dropouts*.

A dropout is simply someone who drops out or withdraws from the study. This might be because he or she dies; suffers from nasty side effects of the treatment; moves house; simply sees no improvement, so that his or her motivation to continue with the treatment is reduced; or for any of the other sorts of reasons we have encountered before. The essence here is that measurements are made over time, or after some elapsed time, and the risk is that the different arms of the trial might be affected by dark data in different ways. They illustrate *DD-Type 7: Changes with Time* and also *DD-Type 1: Data We Know Are Missing*.

For example, in a study comparing an active treatment with a placebo (a "treatment" believed to have no active therapeutic ingredient) any side effects are more likely to occur with the active treatment, since, by definition, the placebo has no active component. This may well mean that dropouts arise more often in the treatment group. Perhaps worse, if people who see no benefit or even deteriorate are more likely to drop out, then the study will be disproportionately left with those who *do* see a benefit. If we don't allow for the dropouts in some way, we can obtain a highly misleading impression of the effectiveness of the treatment. This is another example of survivor bias—those who "survive" or continue to the end of the trial do not properly represent the effect on the entire group.

Things are aggravated—for the good of the individual!—by the need to behave ethically in clinical trials. Point 9 of the Nuremberg Code, established after World War II, requires that participants in clinical trials be able to stop at any time: you cannot force people to stay in a trial.

Trials are often more complicated than the preceding simple two-group illustration. They may involve multiple different clinics and have multiple "arms," with several treatments being compared. An example is given in Figure 4, which shows data from a clinical trial exploring the effectiveness of the drug budesonide in asthma patients.[20] This trial involved five groups of patients, one receiving a placebo (zero dose), and the other groups receiving 200, 400, 800, or 1,600 micrograms of budesonide, respectively. Measures of lung function were taken

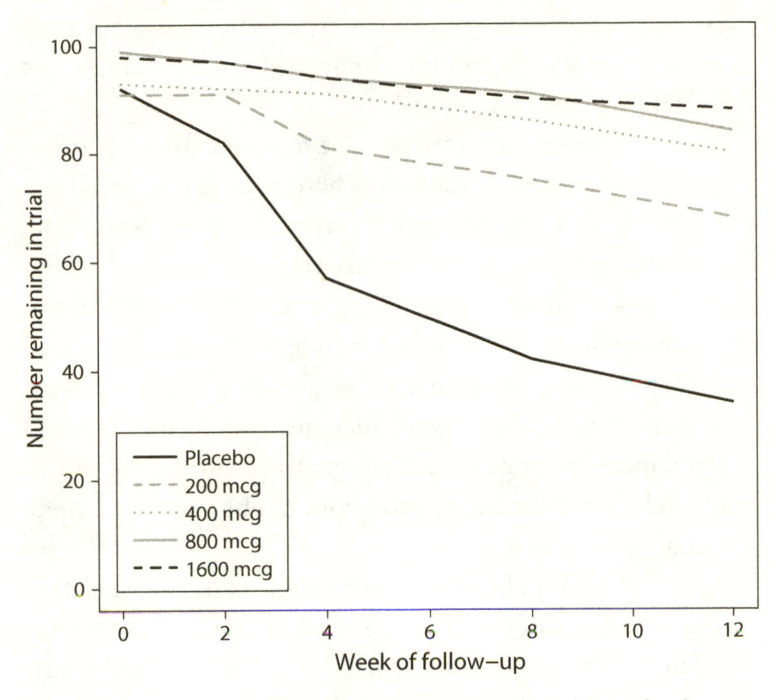

FIGURE 4. Traces showing the number remaining in the asthma trial with time.

at the start, and at weeks 2, 4, 8, and 12 after patients joined the study. The traces in the plot show the number of patients remaining in the study at each week. The others dropped out—and the monotonic pattern is very clear, with there being more dropouts on each measurement occasion. The dropout rate is quite dramatic, with only 75 percent of the patients who began the trial still there at the end. Perhaps even more alarming is that the dropout rate seems to differ among the different arms of the trial. In particular, only 10 of the 98 receiving the highest dose dropped out, whereas 58, that is nearly two-thirds, of those receiving the placebo dropped out. One might speculate that the drug is effective, with only the less severely affected patients remaining in the study, but this would be speculation. Certainly, the missing data complicate the analysis and interpretation of what is going on, with considerable potential to be misled if conclusions are based solely on the data available.

I've illustrated randomized trials using medical examples, but they are also very widely used elsewhere, including in social and public policy domains such as education and crime prevention. Indeed, randomized trials in social and public policy also have a long history—though not so long as in medicine. For example, between 1968 and 1982 researchers in the United States used such methods to investigate whether people would be discouraged from working if they were guaranteed an income equal to the minimum income deemed adequate. (It turned out the income did reduce the time spent working, but only by a small amount.[21])

An interesting application of a randomized trial in education was an evaluation of the impact of the children's TV show *Sesame Street* on vocabulary and cognitive skills. This study illustrated one of the challenges of randomized trials in the social sciences: short of actually forbidding some children to watch the show, there was no way of guaranteeing they wouldn't watch it. The researchers finally got around this problem by finding cities where the show was available only on cable and providing cable to a randomly selected group of households. And they found that watching the show did improve vocabulary. Andrew Leigh describes this and many other two-group randomized trials in his excellent book *Randomistas: How Radical Researchers Are Changing Our World.*[22] He also describes how he set up a randomized trial to find the best title for his book. Before the book was published, 4,000 people each saw a randomly chosen title from one of 12 different possible titles. The response he measured was the proportion of each group that clicked through to the publisher's website for more information.

Regarding crime, it is known that public perceptions of the risk of experiencing crime can exceed the real risk: not knowing something can make it seem worse than it is (another aspect of dark data). A study by the National Policing Improvement Agency in the UK explored whether providing information about crime and policing would be received positively by the public or whether it would increase fear of crime.[23] Four groups of research participants were presented with different kinds of information about crime and policing, with different degrees of dark data. One group received maps showing the recorded crime rate in their local area, another received information about neighborhood policing in their local area, another received both, and a fourth group received no information at all. This last group was completely in the dark. The conclusions were positive: "The study was able to challenge the myth that sharing information with the public would increase the 'fear of crime.' In fact, information was found to improve people's perceptions of their neighbourhood and of the local police."

Of course, the results are not always positive—indeed, one of the most important roles of randomized trials is debunking popular myths or perceived "self-evident" truths which are in fact wrong. And sometimes it takes considerable courage to run a randomized trial in the face of a popular belief that something is true.

For example, while a robust imprisonment policy will reduce crime in the short term, this may not be because the miscreants learn the error of their ways but could simply be that it takes them off the streets. And a prison record is not conducive to fitting back into society and finding regular work on release, so short sentences might not help at all; they might even increase crime in the long term. A proper experiment is needed to investigate the policy. This example also illustrates another of the difficulties of randomized trials in social settings. Few judges—and possibly even fewer members of the public—are likely to agree that sentences should be decided randomly. Nevertheless, such experiments have been carried out, with criminals being randomly selected for sentence reduction.[24]

Even when the results are not positive, careful digging can reveal that there's more to things than meets the eye. Andrew Leigh describes four randomized studies investigating whether providing textbooks to schools improved examination results. None of the four studies showed that the policy did improve results, but the reasons were different in the four cases. In one, the books were put into stores instead of being handed out. In the second, parents compensated for the free books by cutting back their spending on education. In the third, there was no incentive for the teachers to use the books. And in the fourth, the books failed to help the majority of the students simply because they could not read. Had these explanations not been unearthed—had they remained as dark data—the conclusions could have been misleading.

Randomized trials are a wonderful scientific tool, but they are not always appropriate. You would not carry out such a comparative trial to see if having a parachute was more effective than not having a parachute when jumping from an aircraft.[25] Moreover, subtleties can arise. Consider a study aimed at investigating possible ways to reduce unemployment. An intervention might increase my chances of finding a job, but if that means I displace someone else who would otherwise have taken the job, then the intervention will not reduce unemployment. Likewise, an intervention which replaced a few full-time employees by many part-time employees might or might not be regarded as successful even though more people had jobs, depending on how unemployment is defined.

Yet further complications arise from the "Hawthorne effect." This is the tendency people have to behave differently if they know they are being observed. It looks as if, ideally, people should be studied secretly—hiding the fact from them—but this obviously runs into ethical problems of informed consent. The first clause of the Nuremberg Code for clinical trials says that "*The voluntary consent of the human subject is absolutely essential.*"

The general term for the branch of statistics concerned with deciding how to allocate objects to treatments so that analysis can yield the best estimates of treatment effectiveness is simply *experimental design*. The two-group randomized trial is just about the simplest possible and the most widely used design for exploring the effectiveness of a treatment, policy, or intervention. Immediate generalizations of the two-group strategy lead to studies with several groups as illustrated previously, each receiving a different treatment, and to carefully balanced designs in which multiple factors likely to influence the results are taken into account. I describe a plastic car-parts injection-molding experiment like this in chapter 8. Sophisticated designs go under exotic names like fractional factorial and Greco-Latin square designs.

The principles of experimental design were extensively developed by the leading British statistician Sir Ronald Fisher, working on agricultural experiments at the Rothamsted Experimental Station in Hertfordshire, the longest-running agricultural research institution in the world. In 1935, he published a seminal book, *The Design of Experiments*, describing how best to assign "experimental units" to treatments, such as different fertilizers, soil types, irrigation, temperatures, and so on, and how best to explore different treatment combinations. The subject has now become a highly advanced and quite mathematical discipline. It includes strategies such as adaptive allocation, in which the choice of treatment is influenced by results which have been obtained as the experiment proceeds. For example, as time progresses and results begin to accumulate, we might see that they seem to favor a particular treatment. That leads to the question of whether we should therefore assign more units (e.g., people, if we are running a clinical trial) to that treatment on the grounds that it currently appears to be the best or instead allocate them to the other treatments with a view to becoming more confident about the conclusion.

The web makes randomized trials of certain kinds of social intervention very easy to carry out, since we can simply arrange that people are randomly assigned to receive different messages, see different versions of a website, or get different offers—recall Andrew Leigh's experiment investigating the best title for his book. Internet companies typically do this as a matter of course, carrying out thousands of experiments automatically every day so that they can determine the most successful strategy for the company. But the use of dark data in this way can produce a backlash. It inevitably means that customers are being kept in the dark, and if they find out, they may not like it. For example, if randomization is applied to the price charged for goods or services, then having prices change in what appears to be an erratic way can be unsettling and disturbing. In October 2000, the *Washington Post* reported customer reactions when they found out that Amazon had explored the price sensitivity of its customers by randomly manipulating prices.[26] Feedback included: "I thought the idea was to attract customers first and then work hard to keep them. This is definitely not going to earn customer loyalty." And worse: "I will never buy another thing from those guys!"

Randomly varying prices to explore how much customers are prepared to pay is one thing. Other web experiments tread a much finer ethical line. In 2014, Facebook was severely battered by the reaction to an experiment it had carried out to test if "exposure to emotions led people to change their own posting behavior."[27] In the study, Facebook manipulated information posted on nearly 700,000 users' home pages by reducing the amount of positive or negative content, to see if it would make people feel more positive or negative. When the study became public, it was described as scandalous, spooky, disturbing, unethical, and even possibly illegal. It certainly appeared to contravene Article 1 of the Nuremberg Code, that participants should have given consent to be part of the experiment.

Beware Human Frailties

This chapter has looked at three fundamental ways of collecting data. The resulting data strip away obscurity and shine a light to reveal new worlds to us. But the data collection strategies are devised and run by humans, and the resulting data are interpreted and analyzed by humans. Our decisions about what data to collect and what the results of our analyses mean will be based on our previous experience, and this may not be representative of what the world will throw at us in the future. At a deeper level, our decisions will be based on how our evolutionary development has shaped us. One consequence of all these factors is that we make mistakes, we fail to properly balance evidence, and we often fail to take the most rational course. In short, we are vulnerable to all sorts of subconscious biases.

Availability bias is distortion arising from the tendency to judge the likelihood of an event on the basis of how easily we can bring to mind an example of it. If a plane crash has recently been in the news, then there is a tendency to think plane crashes are more likely. Advertisements capitalize on this sort of effect, trying to ensure that when you want to buy a product you most easily recall a particular brand and are less aware or unaware of competing brands. But the effect also at least partly explains sudden increases in rates of diagnosis of particular medical conditions or kinds of social abuse when society focuses on them. For example, in chapter 3 we will look at how the rate of autism diagnosis in the United States has increased since 2000. Part of the explanation for this increase is due to availability bias. If people are bombarded with mentions of a condition, it's hardly surprising that they are alert for indications of it. Indeed, studies have shown that parents living near someone with the condition are more likely to explore such a diagnosis for their own child.

Availability bias is related to the *base rate fallacy*. Suppose you are being tested for a rare disease and know that the test is 100 percent accurate in detecting the disease when someone has it, and 99 percent accurate when someone does not. What is your conclusion if the test is positive, flagging you as having the disease? At first glance you might think it means you are almost certain to be ill. But this can be very misleading, because the correct answer depends on the base rate—on how many people in the population have the disease. For example, if the disease is so rare that only 1 in 10,000 people have it, then, of every 101 people flagged as positive by the test, only 1 person will in fact have it (on average!). Even though the test rarely makes a mistake with people who don't have the disease, because there are so many more of them than there are of people with the disease, almost all those diagnosed as ill are incorrectly diagnosed. If we ignore or are unaware of the fact that the majority in the population don't have the disease, mistakes can be made. In a test on medical students at Harvard, over 56 percent of the students got this wrong. Alarmingly, other studies have shown similar results with practicing doctors. The base rate, at least to these students and doctors, was dark data.

The base rate fallacy arises because people are unaware of or ignore relevant data. A similar phenomenon leads to the *conjunction fallacy*. This is the tendency to regard highly specific conditions as more probable than more general ones. A standard illustration takes the following form.

My friend Fred is a university history professor who teaches courses on Victorian England, nineteenth-century America, and world trade in the nineteenth century. In his spare time he enjoys reading dense biographies and spending his holidays in places of archaeological or historic interest. Now, which do you think is more probable: (i) that Fred has a beard or (ii) that Fred has a beard and is a trustee of his local historical museum?

Many people say (ii), but a moment's thought shows that this cannot be. (ii) is a subset of (i), so (i) must be more likely than (ii). The reason people make this mistake appears to be, much like with availability bias, that they base their conclusion on matching (i) and (ii) to the description of Fred. Someone with Fred's characteristics might be quite likely to be a trustee of his local historical museum.

Confirmation bias is another related risk. Whereas the base rate fallacy and availability bias arise from ignoring data describing the population, in confirmation bias people actively, albeit subconsciously, seek data which do not adequately represent the population. In particular, they seek information which supports their point of view, tending to ignore data not supporting it. Take the example of Jeane Dixon. Ms. Dixon, whose real name was Lydia Emma Pinckert and who died in 1997, was one of America's best-known psychics. She had a widely syndicated newspaper column, and her biography, *A Gift of Prophecy: The*

Phenomenal Jeane Dixon, sold over 3 million copies. In fact, of the vast number of predictions she made, many turned out to be incorrect. But it is possible that she herself believed she had the ability to predict the future, based on recalling the predictions which came true and tending to ignore the others. It's certainly true that many people believed in her powers, again based on noting the occasions when she happened to get them right and paying less attention to the others, effectively casting a shadow over some of the data, turning them dark, a case of *DD-Type 3: Choosing Just Some Cases.* Psychological experiments in which people are presented with a series of examples have demonstrated how what you believe can influence what you remember.

The complementary side of confirmation bias is that people also have a tendency to *forget* disproving evidence if it is contrary to their initial beliefs.

Other examples of how people draw inappropriate conclusions because they ignore part of the data (probably subconsciously) include *negativity bias,* in which people have a natural tendency to more easily recall unpleasant events than pleasant events; *acquiescence bias,* in which respondents say what they believe the interviewer wishes to hear; the *bandwagon effect,* in which people tend to go along with the majority; the *belief bias,* in which the particular response given depends on how believable the respondent believes it to be; and the *bizarreness effect,* in which striking material is remembered better than common material. All in all, it seems a wonder that we ever get anything right!

Not surprisingly, the sort of phenomena described in this section often lead to overconfidence: if most of the evidence you can recall supports a particular position, then you might reasonably be very confident that that position is right. And the problem is aggravated by so-called echo chambers, which we discuss in chapter 5.

You might think that once you have been made aware of these human frailties you can avoid them. That might be true to some extent, but they have a tendency to sneak up on you unexpectedly. This is illustrated by studies showing how the wording of survey questions can lead to inconsistent answers. A familiar example arises when both positive and negative versions of a question are asked (e.g., "Did you like the film?" versus "Did you dislike the film?"). In principle (assuming there is no "no opinion" box) the numbers of people answering no to the first and yes to the second should be equal. But often they aren't. Some kind of measurement error, buried deep in the minds of those answering the question, is serving to conceal the true value.

Chapter 3

DEFINITIONS AND DARK DATA

What Do You Want To Know?

It may be obvious, but it's nevertheless worth emphasizing, that the usefulness of data depends on the right data having been collected in the first place and on the data being collected without distortion or misrepresentation. Each of these conditions is vulnerable to dark data risks. In fact, each of them is subject to so many potential dark data risks that an exhaustive enumeration of the relevant risks would be impossible. Nonetheless, developing an awareness of the sort of situations to look out for is vital in coping with dark data. This chapter examines what data you aim to collect, and the next looks at how well that is achieved, in both cases from the perspective of the dark data risks.

Different Definitions and Measuring the Wrong Thing

A fundamental kind of dark data arises from using inappropriate definitions—or, as we might say, not knowing what you are talking about. Let's look at some examples.

Immigration

Surveys are specifically designed with the target questions in mind, but administrative data might have been collected for some entirely different reason. This means that administrative data might not be suitable for answering the question you are interested in. For example, a recent controversy arose in the UK about the accuracy of Long-Term International Migration (LTIM) statistics. The Office for National Statistics (ONS) gave a figure, based on the International Passenger Survey (IPS), of 257,000 for the number of people immigrating to the UK from the European Union for the year ending September 2015. However, the number of European Union nationals registering for National Insurance Numbers (NINos) over this period was 655,000. NINos are personal account numbers for people who work in the UK, to ensure that tax payments and National Insurance contributions (e.g., for health services and pensions) are properly recorded, so the discrepancy appears odd to say the least. It looks as if the ONS numbers are way off. In fact, UK politician Nigel Farage said: "They are pulling the wool over our eyes. NINos are a simple and clear reflection of the real numbers of people in this country, as without them you can neither legally work, nor claim benefits."[1]

The IPS, which covers all major air, sea, and tunnel ports into the UK, has been running continuously since 1961. It carries out between 700,000 and 800,000 interviews each year. Although that number represents just a fraction of the people traveling to and from the UK, the responses can be used to estimate the total number of migrants. But it is just an estimate, and there is inevitably some uncertainty associated with it. The ONS actually gave a measure of this uncertainty of ±23,000, yielding an interval of 234,000 to 280,000, saying it was 95 percent confident the range contained the true number. The uncertainty, while substantial, clearly does not explain the difference from the NINo figure.

Consequently, the ONS explored the discrepancy between this estimate and the NINo number in depth.[2] It found that short-term migration ("migrants who arrive for . . . between 1 to 12 months") was the main reason for the difference. Long-term migrants stay for 12 months or more. Short-term migrants can work and apply for NINos, but LTIM (long-term) numbers are the headline figure. The ONS went so far as to comment that "definitional differences between these data are fundamental, and it is not possible to provide an accounting type reconciliation that simply 'adds' and 'subtracts' different elements of the NINo registrations to match the LTIM definitions. . . . NINo registrations data are not a good measure of LTIM. . . ." In short, the administrative data were aimed at the operation for which they were collected and might not be ideal for other purposes. Inappropriate or unsuitable definitions effectively obscure the data of interest. They illustrate dark data *DD-Type 8: Definitions of Data,* and the key point to remember is that whether data are dark or not depends on what you want to know.

Crime

Another example of dark data arising from differences in definitions is illustrated by crime statistics. At a national level, crime statistics in England and Wales come from two main and rather different sources: the Crime Survey for England and Wales (CSE&W) and the Police Recorded Crime (PRC) statistics. The CSE&W is equivalent to the National Crime Victimization Survey in the United States. It was launched in 1982 (as the British Crime Survey), asking people about their experience of crime over the preceding year. The PRC data are collated from the 43 police forces of England and Wales and the British Transport Police, and are analyzed by the Office for National Statistics.

The different nature of these two processes of collecting information has immediate dark data implications. By definition, since the CSE&W survey asks people what crimes they have experienced as victims, it does not report murders or possession of drugs. It also does not include people who live in group residences, such as care homes or student residences, and it does not cover crime against commercial organizations or public bodies. As you can see, there is a large potential for dark data, though this risk is made explicit by having a clear definition of what is covered by the survey.

The PRC statistics also have dark data, though of a shade complementary to that of the CSE&W. By definition, the PRC statistics do not include crimes not reported to the police, perhaps because the victim thought the police would do nothing. This matters, because it has been estimated that only about 4 in 10 crimes are reported, though obviously this number varies substantially among types of crime. Moreover, the crimes appearing in the police statistics are those which are classified as one of a number of "notifiable offences," these being offenses that could possibly be tried by jury (and a few others). Further complications arise from feedback mechanisms (*DD-Type 11: Feedback and Gaming*). For example, the number of drug possession offenses will depend on the extent of police activity, and the extent of police activity might depend on perceived rate of drug possession, which in turn will be influenced by the number of drug possession offenses in the past.

The use of different definitions explains discrepancies between the crime rates reported by the two sources. For example, in 1997 the PRC recorded 4.6 million offenses while the CSE&W

estimated there were 16.5 million. The differences also explain something which has been very puzzling to media pundits and casual readers, namely, that, according to the PRC, the numbers of crimes went up from 1997 to 2003 (from 4.6 million to 5.5 million) while according to the crime survey the numbers went down (from 16.5 million to 12.4 million).[3] So, is crime going up or down? You can probably guess which side the media chose to highlight.

Medicine

Immigration and crime are just two of an unlimited number of domains in which definitions can lead to dark data by failing to include cases or types we might have wanted to include. Sometimes the consequences can be surprising. For example, dark data issues of definition can explain why more people are dying of conditions related to Alzheimer's disease than in the past.

Alzheimer's disease is the most common form of dementia. It is progressive, with early stages typically involving slight memory loss, and later stages characterized by confusion, an inability to understand what's going on, and personality changes. It is thought to affect some 50 million people worldwide, a number which is increasing and forecast to reach 75 million by 2030. And dark data can explain this increase in at least two ways.

First, no one died of it before 1901, since that was when the German psychiatrist Alois Alzheimer first described a case of the disease which was subsequently named after him. Moreover, the diagnosis was originally reserved for people aged between 45 and 65 with symptoms of dementia. Only later, in the last quarter of the twentieth century, was the age range restriction relaxed. Clearly, such broadening of definitions will change the numbers of people being diagnosed with the disease. Data previously treated as not relevant are becoming visible.

The second way in which dark data explain why more people are dying of conditions related to Alzheimer's disease than in the past might appear paradoxical: it arises because medical science advances. Thanks to medical progress, people who would have died young are now surviving long enough to be vulnerable to potentially drawn-out deteriorating conditions like Alzheimer's. That raises all sorts of interesting issues, not least of which is the question of whether extending life is necessarily beneficial.

The fact that the rate of diagnosis of autism in the United States has doubled since the year 2000 can also be largely explained by dark data *DD-Type 8: Definitions of Data*.[4] We saw in chapter 2 that one reason for the increase was availability bias—awareness of the condition was being raised. But another very important reason for this increase is that there have been changes to the formal *way* autism is defined and diagnosed. In particular, although autism was included in the *Diagnostic and Statistical Manual of Mental Disorders* in 1980, how it was diagnosed was changed in 1987 and in 1994 in ways which essentially relaxed the criteria. Relaxing diagnostic criteria, so that you are making them easier to meet, means that more people satisfy them.

Furthermore, in 1991 the U.S. Department of Education decided that an autism diagnosis qualified a child for special educational services, and in 2006 the American Academy of Pediatrics recommended that all children be screened for autism during routine pediatric visits. If you change the way data are used, it is perhaps hardly surprising that behavior in collecting those data changes—a feedback phenomenon of the kind we examine in detail in chapter 5. The same sort of effect is also illustrated by the consequences of the launch in England in February 2009 of campaigns for national dementia awareness along with a National Dementia Strategy aimed at improving diagnosis rates and the quality of care. Perhaps hardly surprisingly, the rate of dementia diagnosis increased, going up by an estimated 4 percent in 2010 and 12 percent in 2011, compared with 2009.[5]

Economics

In general, changing definitions over time can obviously lead to changes in the nature of the data which are collected. Not only can this make retrospective comparisons difficult, but it can also lead to accusations of dishonesty. An obvious example is given by definitions of unemployment: change the definition, and a government's performance can suddenly look much better.

Another example in economics is the measurement of inflation. Definitions of inflation are based on recording the prices of a defined set of goods and services (called a "basket" of goods and services—not a literal basket, of course) and seeing how the average price changes over time. However, there are various complications—all hinging on dark data *DD-Type 8: Definitions of Data*. One is the question of how to calculate the average, since statisticians have several different ways to do this—the arithmetic mean, geometric mean, harmonic mean, and so on. Recently, the UK switched from using an index based on the arithmetic mean to one based on the geometric mean, bringing it into line with most other countries. Using a different definition means you are looking at things from a different perspective, so that naturally you see—and don't see—different aspects of the data.

Beyond any effect originating from changing the formula, dark data also arise in a more fundamental way in inflation indexes: it is necessary to decide what items to include in the basket and how to obtain their prices. In general, as the earlier examples showed, we must be aware of the risk of dark data whenever a choice is made in a data collection process. Here, the question of what to put in the basket is potentially problematic, because society changes, and inflation indexes are supposed to reflect the cost of living in some way. That last phrase is deliberately ambiguous—"in some way"—because different indexes measure different aspects of inflation experience. Some measure how price changes affect individuals, others how they affect the larger economy, and so on. In any case, it is important that the basket of items be *relevant*—that is, it must consist of goods and services that people actually buy. The challenge is illustrated clearly by comparing what might have been included in a price index basket 200 years ago with what might be included today. Two hundred years ago, candles would have been an important item to include in the basket, but candles generally do not constitute much of people's expenditure today. Instead, cellphones and cars will figure. This means that we have a nominal list of items which could in principle be included in the basket, but we will not want to include all of them. Although a great amount of thought goes into specifying exactly what items should have their prices recorded, it is clear that there is scope for ambiguity and, indeed, arbitrariness to creep in.

As to the second point, how to obtain the prices for the items in the basket, the traditional way has been to carry out surveys and send teams of researchers out to shops and markets to note the prices of goods. The U.S. Bureau of Labor Statistics surveys some 23,000 businesses and notes the prices of some 80,000 consumer items each month, aggregating these to yield the Consumer Price Index. Other countries carry out similar exercises.

You may have spotted that this traditional approach to collecting the prices of goods totally ignores online shopping. Given that such purchases now account for around 17 percent of UK retail sales[6] and nearly 10 percent of U.S. retail sales,[7] it seems that a large number of relevant prices may not contribute to the index. (I should add that those figures are "at the time of writing," since the trends are upward and steep.) For this reason, many countries are also developing measures based on web-scraping of online prices. The measures do not seek to precisely replicate the traditional measures, because the baskets will differ. We will see an example of such a measure in chapter 10.

Society is constantly changing—perhaps more so now than at any time in the past, as the computer and its attendant technologies such as surveillance, data mining, artificial intelligence, automated transactions, and the web continue to have an impact. Such a rapid rate of change has important general implications for data analysis in terms of dark data, because projections about the future are necessarily built on what happened in the past. In technical terms, sequences of data over time are called, naturally enough, *time series* of data. The rate of change of data collection methods and technologies means that the time series we need often do not go back very far. New types of data necessarily have short histories, so data will be available from only the relatively immediate past. Beyond that lies darkness.

You Can't Measure Everything

Data sets are always finite. This is certainly true in terms of numbers of cases—the finite number of people in a population or the finite number of times something is measured. But it is also the case in terms of *what* is measured or *what data* are collected about the objects of interest. If studying humans, we might determine their age, weight, height, qualifications, favorite food, income, and a host of other things. But there will always remain countless other characteristics we have not determined. These other characteristics are inevitable dark data, with consequent implications.

Causality

When population studies suggested a link between lung cancer and smoking, the eminent statistician Ronald Fisher pointed out that this did not necessarily mean that smoking caused cancer. Among other possibilities, he noted that perhaps both lung cancer and a predisposition to smoke were caused by some other factor—maybe, for example, a genetic condition which encouraged both things. This would be a classic case of dark data *DD-Type 5: Missing What Matters*—some other, unmeasured variable which causes both and so induces a correlation between them, even though neither causes the other. It also shows how difficult it can be to detect dark data.

In fact we encountered this sort of situation at the start of the book. I mentioned in chapter 1 that for children in their early years of schooling, height is correlated with vocabulary. So, if you conducted a survey measuring the height and testing the vocabulary sizes of a sample of children aged between 5 and 10, you would find that the taller children had, on average, a larger vocabulary than the shorter children. You might conclude from this that teaching children more words would make them grow taller. And, indeed, if you did carry out such an exercise, measuring the initial heights of a group of children, giving them intensive lessons in new words, and then measuring their heights again at the end of the year, you would find that their height had increased.

But the reader will know what's going on here. While height and vocabulary for such children are certainly correlated, it is not because there is a causal relationship between them. The fact is that both are related to a third variable, perhaps one we didn't think to measure in our survey, namely, the ages of the children. Age was a dark data variable in this study, and failure to measure it could give a very misleading impression of what the data were showing.

This situation differs from ones in which some people (or, more generally, objects) are missing values on some attributes in a record, and it differs from ones in which *all* the attributes are unrecorded for some people (or objects). Now, the values for a certain attribute or certain attributes are missing for all *cases* in the database. All entries for such a variable would be recorded as blanks or NA, if the variable was noted at all. For example, in a survey, perhaps an oversight meant we simply forgot to include the question asking how old the respondents were, so we do not have age information for anyone. Or perhaps we didn't think age would be relevant, so it never occurred to us to include that question in the first place. None of this is at all far-fetched: if surveys are too lengthy, there will be an adverse impact on response rate, so careful choices have to be made about what questions to include.

Paradox!

Sometimes *DD-Type 5: Missing What Matters,* in which entire variables or characteristics are missing from the data, can have some rather puzzling implications.

Everyone is familiar with the tragedy of the *Titanic*—the unsinkable liner that sank. But close examination of the survival rates among the passengers and crew reveals something curious.[8] As Table 2(a) shows, there were 908 crew members on the ship, and of these just 212 survived, that is, 23.3 percent. And among

TABLE 2. Proportion of crew and third-class passengers who survived the sinking of the *Titanic*: (a) overall; (b) men and women separately

(a)

Crew	Third-class passengers
212/908 = **23.3%**	151/627 = **24.1%**

(b)

	Crew	Third-class passengers
Men	192/885 = **21.7%**	75/462 = **16.2%**
Women	20/23 = **87.0%**	76/165 = **46.1%**

the 627 third-class passengers—those traveling deepest in the ship and who therefore found it most difficult to get out—just 151 survived, that is, 24.1 percent. Although there's not much difference between the survival rates of these two groups, it shows that passengers were slightly more likely to survive than crew.

But now let's look at the survival rates for men and women separately, shown in Table 2(b).

First, for the men. Among the crew there were 885 men, of whom 192 survived, a rate of 21.7 percent. And among the third-class passengers 462 were men, and 75 survived, a rate of 16.2 percent, so male crew had a higher survival rate than male third-class passengers.

Second, for the women. Among the crew there were 23 women, and 20 of those survived, a rate of 87.0 percent. And among the third-class passengers 165 were women, and 76 survived, a rate of 46.1 percent, so female crew had a higher survival rate than female third-class passengers.

So what is going on? For the men and women separately, the crew had a higher survival rate than the third-class passengers.

But overall the crew had a lower survival rate than the third-class passengers.

This is not a trick—the numbers are what they are. But it does seem almost paradoxical, and in fact the phenomenon is often called *Simpson's paradox*, after Edward H. Simpson, who described it in a paper in 1951 (though others had also described the phenomenon at least 50 years earlier).

The implications are potentially serious. If we had not recorded the sex of the people on the ship—if those data had been missing—we would quite happily have reported the results of our analysis to be that third-class passengers were more likely to survive than crew. But that would be misleading if we were interested in males—since the results for them are the opposite. And likewise, it would be misleading if we were interested in females. That means the conclusion would be misleading if we were interested in *anyone*, since each of the passengers was male or female.

We will explore why this situation occurs in a moment, but quite clearly the potential consequences of it are staggering. An unlimited number of characteristics of people who sailed on the *Titanic* weren't recorded. If any of them could have the effect of flipping our conclusions, then omitting them, and their being missing data, could be very misleading. This might not matter in the case of the *Titanic*, since in that case we are merely describing historical data, but consider the following example.

Suppose we are carrying out a clinical trial of the kind discussed in the previous chapter, comparing drug A with drug B. To compare the drugs, we give A to one group of people and B to another group. Both groups contain people of various ages, which for convenience we will name "younger" and "older," according to whether they are younger or older than 40, say. To make things concrete we will suppose that the group receiving

TABLE 3. Average scores for drugs A and B: (a) for younger and older separately; (b) overall

(a)

	Average score	
	Drug A	Drug B
Younger	8	6
Older	4	2

(b)

Average score	
Drug A	Drug B
4.4	5.6

drug A has 10 younger and 90 older in it, while the group receiving drug B has 90 younger and 10 older.

Now let's look at the results, where we suppose that the higher the score the more effective the treatment. These (hypothetical) results are given in Table 3.

Suppose we find the average score for the younger in group A is 8, and the average score for the younger in group B is 6, as shown in Table 3(a). This tells us that drug A is more effective for the younger, since 8 is greater than 6.

Likewise, for the older suppose we find the average score in group A is 4, and the average score in group B is 2, as in the second row of Table 3(a). For the older, drug A is more effective than drug B.

Although the average score for the older is lower than that for the younger whichever treatment is given, it is clear that, for both the younger and the older, drug A is more effective than drug B. We should recommend that drug A be prescribed.

But what about overall? The overall average score of all people receiving drug A is $(8 \times 10 + 4 \times 90)/100 = 4.4$, while the overall average score of all people receiving drug B is $(6 \times 90 + 2 \times 10)/100 = 5.6$. These results are shown in Table 3(b). Overall, when we ignore the age of the patients, drug B produces a higher score than drug A.

This means that if we had not recorded the age of the patients—if those data were missing—then we would be led to the conclusion that B was better than A, even though for the younger A is better than B, and for the older A is better than B, that is, even though *for everyone* A is better than B.

An obvious initial response is that we should record age while collecting the data. That's all very well, but once again there are countless other variables we might also record, any of which could have this same strange inversion effect. And we cannot record *all possible* variables. Some are inevitably dark data.

The key to the mystery lies in how those overall averages are calculated. In the drug trial example, in group A there are far more older than younger, while in group B the reverse is true. That swings the overall averages down: 8 is bigger than 6, and 4 is bigger than 2, but if you put enough weight on the 4 when you calculate the average of 8 and 4, and put enough weight on the 6 when you average the 6 and 2, things turn around.

So we see now what caused the problem—it was the difference between the proportions of younger in the two groups. The group receiving drug A had 10 percent younger, while the group receiving drug B had 90 percent younger. If the two groups had had equal proportions of younger, then the problem would not have arisen. Since the drug trial is an experiment, in which we control how many patients receive each treatment, we could have eliminated the problem by balancing the proportions of the younger so they were the same in each group.

That method works if we have control over who goes into which group. But there was no such control in the *Titanic* example—passengers were passengers, and crew were crew, and we did not choose which was which.

The following is another example in which we don't have control over who is in which group.

In a 1991 study of the effect of race on death-penalty sentences in murder convictions in Florida, 53 of 483 Caucasian defendants and 15 of 191 African American defendants were sentenced to death;[9] that is, a higher proportion of Caucasians (11.0 percent) than African Americans (7.9 percent) were sentenced to death, as shown in Table 4(a).

But if we now take the race of the *victim* into account, as well as that of the defendant, a rather different, and again puzzling, picture emerges.

As is shown in Table 4(b), for cases in which the victim was Caucasian, we find 53 out of 467 (11.3 percent) of the Caucasian defendants sentenced to death, and 11 of the 48 African American defendants (22.9 percent) sentenced to death. And for cases

TABLE 4. Death penalty proportions: (a) overall; (b) by race of victim

(a)

Defendant	
Caucasian	African American
53/483 = **11.0%**	15/191 = **7.9%**

(b)

		Defendant	
		Caucasian	African American
Victim	Caucasian	53/467 = **11.3%**	11/48 = **22.9%**
	African American	0/16 = **0.0%**	4/143 = **2.8%**

in which the victim was African American, we find 0 out of 16 (0 percent) of the Caucasian defendants sentenced to death, and 4 of the 143 African American defendants (2.8 percent) sentenced to death. So, it seems that when the victim is Caucasian, a higher proportion of African American than Caucasian defendants (22.9 vs. 11.3 percent) receive the death penalty; and, *also*, when the victim is African American, a higher proportion of African American than Caucasian defendants (2.9 vs. 0 percent) receive the death penalty. But overall, a lower proportion of African American than Caucasian defendants (7.9 vs. 11.0 percent) receive the death penalty.

As before, the explanation lies in the imbalance in the proportions between the groups. The overall proportion for Caucasian defendants (11.0 percent) arises by averaging outcomes for 467 Caucasian victims and 16 African American victims. But the overall proportion for African American defendants (7.9 percent) arises by averaging outcomes for 48 Caucasian victims and 143 African American victims. These two ratios, 467:16 and 48:143, go in opposite directions, and that distorts the overall averages.

Once again you might legitimately say: "OK, the analyses yield different results, and we can see why. *But both seem reasonable, so which is right?*"

The answer is, it depends on the question. In particular, do you want to ask a question about the overall population, with its relative group sizes as given, or do you want to make a comparison within groups? If the former, then it's OK to ignore the splitting variable. If the latter, then obviously it's necessary to include it.

Note that the drug trial example is a little different from the other two. In situations like in this example, there would normally be nothing fixed and preordained about the number of younger and older in each group: these numbers would be chosen by the experimenter. In contrast, in the *Titanic* and death penalty examples, the numbers were exactly as they were in the populations considered—the people on the *Titanic* and those sentenced to death. Thus, in those two examples it makes sense to talk about the entire populations, whereas in the drug trial example, since the experimenter chose the mix of younger and older and could have chosen a different mix, it probably doesn't. (There might be cases in which the aim is to see how effective a treatment is on a population as a whole, with the mix of younger and older as it occurs in that population.)

All this means, first, is that it's necessary to be very clear about what question you are asking, and, second, that whether data are dark or not will depend on that question. Trite though it may sound, the data you need to collect, the analysis you will undertake, and the answer you will get depend *on what you want to know*.

Between or within Groups?

Issues similar to Simpson's paradox also occur in other guises. For example, the *ecological fallacy* describes a contradiction between an overall correlation and a correlation within groups of objects. A classic example was described in 1950 by the sociologist W. S. Robinson.[10] The correlation between the percentage of the population who were foreign-born and the percentage who were literate over the 48 states of the United States in 1930 was 0.53. That means that states which had a higher proportion of foreign-born were also more likely to have a greater literacy rate (more people could read, American English at least). Superficially, this value suggests that being foreign-born meant people were more likely to be literate. But looking within states revealed a rather different picture. Within states the average correlation was −0.11. The negative value means that being foreign-born meant people were *less* likely to be literate. Had the within-state information not been available—had it been dark data—an incorrect conclusion could have been drawn about the relationship between birth country and literacy.

A related problem known as *omitted variable bias* arises in more sophisticated statistical techniques. *Multiple regression* is a statistical tool which relates multiple predictor variables to a response variable. In general, though there are special cases in which this does not apply; omitting any of the predictors will lead to different apparent relationships between the other predictors and the response variable. And, as before, since we cannot include *all possible* predictors, we will inevitably be omitting some. The risk arising from the absent dark data is that the conclusions may be misleading. Of course, statisticians are aware of these issues and have developed tools to try to mitigate them.

Screening

Early in my statistical career I took part in a project to identify women who were likely to be suffering from osteoporosis. Osteoporosis is a condition in which the bones lose mass and become fragile, so that they are more likely to fracture—one reason why falls in the elderly can be so dangerous. There are sophisticated ways to measure bone density, using things like central dual-energy X-ray absorptiometry (central DXA), but these tend to be elaborate and relatively expensive. This means that it is desirable to develop a screening process allowing the identification of those people likely to be suffering from the condition. In the work I was involved in, the aim was to develop a simple pen-and-paper questionnaire which would produce a score of the likelihood of having osteoporosis based on known risk factors. This could be filled in by hand, without requiring an expert to operate a machine.

Our questionnaire, like other screening instruments, would not perfectly separate those with the condition from those without it. But even an imperfect measure which could separate the classes reasonably, identifying most of those at high risk and most of those at low risk, would be immensely valuable. It would allow the health services to focus attention on those most likely to have the condition, testing them with the more accurate and expensive procedure.

Now, an imperfect system can make two kinds of error. On the one hand it might fail to identify someone who has the condition—in the osteoporosis example perhaps someone who has it but has none of the standard risk factors. And on the other hand it might flag someone as likely to have the condition even though he or she does not. Obviously, the lower the rates of these two types of error, the better the screening instrument will be. It is straightforward enough to reduce the rate of the first type of error to zero by simply classifying everyone as having the condition, but this would make the screening exercise pointless. Worse still, it would mean the other type of error would be large: we would be classifying everyone without the condition as having it. Likewise, we could reduce the rate of the second kind of error to zero by classifying everyone as condition-free, but this would be just as pointless. Given that the instrument is not perfect, we must strike some kind of balance. Or, to put it another way, we must accept that we will misclassify some people.

Anyone the screening instrument decides is likely to have the condition will be given a closer examination. In the osteoporosis example, perhaps they will go on to be tested using central DXA. These people will include those without the condition that we misclassify as having it—so that we will discover that they do not have it. But the others, those that the screening instrument classifies as not having the condition, will not be investigated more closely. We will not learn which of those we got wrong, or at least, not until their condition has progressed further. We might hope (if the screening instrument is reasonably effective) that there are not too many of them, but their true condition will be dark data.

Misclassifying persons as healthy when they have the condition can be serious, especially in the case of a potentially fatal but easily treatable condition. But misclassifying someone as having the condition when he or she does not can also be unfortunate. For example, suggesting that someone might have a serious illness such as AIDS or cancer can obviously have adverse psychological consequences, even if it later turns out to be a mistake. It will also be associated with the unnecessary costs of carrying out the more accurate examination (the DXA examination in the osteoporosis example). Gerd Gigerenzer, an expert on how people misunderstand chance and statistics, gives the example of breast cancer screening programs.[11] He points out that of 1,000 women who take part in such screening programs about a hundred are incorrectly flagged as meriting further investigation, thus being exposed to intrusive, possibly painful, or disturbing examinations. Indeed, even for those in whom breast cancer is detected the outcome might be worse than if it had not been detected. As Gigerenzer puts it: "Women who do have breast cancer, but a nonprogressive or slowly growing form that they would never have noticed during their lifetimes, often undergo lumpectomy, mastectomy, toxic chemotherapy or other interventions that have no benefit for them." Sometimes, it seems, it might be better if the data remain dark.

The measurement of the effectiveness of screening programs is complicated by progress over time. For example, as we saw earlier, the rise in the prevalence of Alzheimer's disease is at least partly a result of people living longer, so that an Alzheimer's diagnosis is no longer invisible dark data about "what might have happened had you survived long enough to suffer from it" but is becoming visible data about "what actually happened to you."

Screening programs are also susceptible to a subtle manifestation of dark data called *length-time bias*. The following is an artificial example to illustrate.

Suppose there are two types of disease, one which lasts one day and one which lasts one year, and during the course of which the infected people live normal lives but at the end of either period they die. For simplicity, suppose also that one new person is infected by each of the diseases each day (I said it was an artificial example). If we want to know how many people suffer from each of these diseases, a simplistic (and mistaken!) approach will be to choose a single day and see how many people have each of the diseases. We will find just one person with the shorter-duration disease—the person infected on that day. But we will find 365 people with the longer-duration disease—everyone who was infected up to a year before that day. Superficially, it will look as if 365 times as many people are contracting the longer-duration disease than the shorter-duration disease. What we will be missing will be the other 364 people infected with the shorter-duration disease during the preceding year.

This may seem a contrived and artificial example, but it is exactly what can happen in cancer screening. Slowly growing cancer tumors tend to have a longer presymptomatic phase, and they also have a longer survival time. A study such as that just described would identify more people with slowly growing cancers than with faster-growing cancers. It would lead to a very misleading impression of the rate at which the two cancers strike people in the population.

You can think of screening programs as seeking to assign people to the correct class (such as disease vs. no disease), but the same abstract structure describes many other situations. We previously examined the case of credit scoring, in which the aim was to classify people according to whether or not they would default on repayments for a loan. Personnel selection—choosing people for jobs—provides another example. Many applicants apply, and these are whittled down based on an initial perusal of curricula vitae and application forms. Candidates on a shortlist are then invited for interview. The initial perusal serves the same role as the screening instrument. Candidates interviewed but not offered jobs might be regarded as the false positives—the ones who looked suitable on the basis of the CVs but who, on closer examination, were found to be not among the most suitable. But there will also be a wealth of candidates rejected on the basis of the CVs who would have been perfectly suitable. In medical parlance these are the false negatives—the dark data.

Selection on Basis of Past Performance

We might choose a courier to deliver our packages on the basis that the company has been the quickest in the past. We might choose a car model because of its safety record to date. We might choose a restaurant because we really enjoyed our meals when we previously ate there. It makes sense to predict likely future performance from past performance. Indeed, we often don't have much else to go on. Unfortunately, however, the past can be an uncertain guide to the future. This is obviously the case when things change—if the economy takes a turn for the worse, or

TABLE 5. Change in wheat yield in 1973 relative to 1970 at 12 UK agricultural sites

		1970 yield	
		Low	High
Difference between 1970 and 1973	**Increase**	5	1
	Decrease	1	5

the car manufacturer introduces a new model, or the restaurant changes hands. But performance can also deteriorate if *nothing* changes. In fact, it turns out we should *expect* such deterioration.

This strange phenomenon—leading to an expectation that good performance should deteriorate and bad performance get better even if nothing about the process has changed—is called *regression to the mean*. It's a manifestation of *DD-Type 3: Choosing Just Some Cases*. Once you are aware of its existence, you will see examples of it everywhere. The following is an illustration.

Wheat yields in kilograms were recorded at 12 agricultural sites in the UK in 1970 and 1973.[12] Table 5 shows whether the yield increased or decreased between those years. We see that for the six sites with the lowest yield in 1970, five of them had a larger yield in 1973. Conversely, for the six sites with the highest yield in 1970, five of those showed a decrease in 1973. The pattern is clear and striking, and it means that had we selected just those sites with high 1970 yields in the expectation of similar yields in 1973, we would have been disappointed.

To see why this sort of pattern emerges, suppose we have a group of students, all of the same ability and industriousness. Although they are of the same ability, whenever they take a test some will do better than others because of variability arising from chance conditions—whether they slept well the night before, or their mind was occupied with something else, or they predicted poorly what might come up on the test, and so on. We might rank them according to their test scores and classify the top-scoring student as best.

But what is likely to happen on the next test?

Since we supposed that the students were all of the same ability, the students scoring highest on the first test obtained these scores because of chance circumstances favoring them. These circumstances—and the less fortunate circumstances of the other students—are unlikely to repeat themselves. That means it is likely that the best-performing students will probably do less well in the next test, and the worst-performing students will probably do better.

The problem here is that the initial test results—the past data—show us the natural ability of the student *combined* with the chance influences. The true abilities are obscured by randomness.

Of course, in real life we are unlikely to encounter a group of students who have equal ability and industriousness. They are likely to have a range of abilities. But even so, those students who do particularly well in the first test are likely to do less well in the second, because at least some of their excellent first test performance is likely to be due to luck. The consequence is that if we simply select the best-performing students—for a job or a graduate program, for example—we are likely to find that their performance will not be so good in the future.

What practical message should we take from this example? Does it mean that we should avoid those who have performed best in the past? The answer is generally no, since they may still do well in the future—just not as well as it seemed they would judging from their past performance. In general, the amount of deterioration (or improvement of those who previously did poorly) will depend on the magnitude of the random aspect of the measurement compared with the magnitude of the real ability. If the range of uncertainty arising from the random part is large compared with the range of abilities, the effect will be very pronounced. Note that the uncertainty cannot be seen—all we see is the combination of uncertainty and ability—so both the uncertainty and the true abilities are forms of dark data.

The term "regression to the mean" was coined by the Victorian polymath Francis Galton. He noticed that (on average) the children of tall parents, while still taller than average, were not as tall as their parents, and the children of short parents, while still shorter than average, were not as short as their parents.

In this chapter we have looked at just some of the ways in which we can be misled by unsuspected dark data if we are not careful enough or do not think carefully enough about what we really want to know. In the next chapter we look at how we can be misled even if we are sure about what we want to know.

Chapter 4

UNINTENTIONAL DARK DATA

Saying One Thing, Doing Another

The Big Picture

Measurements cannot be infinitely accurate. Counts, such as the number of children in a family or even of ships on the sea, occur in convenient whole-integer numbers, but measurements like length have to be rounded to some level. They might be rounded to the nearest centimeter, millimeter, micron (a millionth of a meter), tenth of a micron, and so on, but it is not possible to take an *infinite* number of decimal places. Put another way, at some level we are unable to determine the details and have to live with the big picture (even though "big" might be pretty small!). And that, inevitably, means the details are dark.

This "rounding" is apparent whenever you look at tables of data, where you will see numbers like 70.3, or 0.04, or 41.325, or sometimes values like 76.2 ± 0.2, where the ± 0.2 indicates a range of accuracy within which the underlying true number is thought to lie. This last notation makes very explicit the fact that we are facing dark data.

Rounding is necessary and so familiar that often we simply don't notice it is hiding data. For example, people's ages are often recorded to the nearest year despite the fact that ages are not really just whole numbers of years but include number of days—and hours, minutes, and so on, though at some point the intrinsic inaccuracy arising from the fact that a birth takes place over a period of time rather than at a precise instant means that no further precision can be attained. The convention of rounding *down* to the nearest whole number of years when giving one's age means that ages appear to be grouped around whole numbers and that stated ages are smaller than the actual time people have lived.

Sometimes ages are rounded to the nearest five years, or categorized as simply younger, middle, or older, with division points of 25 and 65 years. While this last grouping might be perfectly adequate for some purposes, it's obvious that it loses information: it conceals what is going on *within* the age groups. The problem is most clearly illustrated if we go to an extreme and group people into just two categories, which we might call younger and older, according to whether they are younger or older than, say, 35. From data summarized in this way we can see if the older group has properties different from those of the younger group: for example, we could see if the average income of the older group was greater than that of the younger, or if the proportion married was greater. But we lose the ability to detect more subtle relationships. For example, we could not tell if average income increases from a very young age to some maximum in middle age and then begins to decrease again for older people. The darkening or "coarsening" of the data has denied us the possibility of such potential discoveries, drawing a curtain across our vision.

When data are collected directly by humans, the darkening of data implicit in rounding can occur in a particularly insidious way. It can even lead to mistaken decisions and actions. Simon de Lusignan and his colleagues studied 85,000 blood pressure values.[1] There is no conceivable reason why the last digit of any of these values should be more common than any other in real underlying blood pressure values. That is, we should expect about 10 percent of the values to end in 0, and 10 percent to end in 1, and 10 percent to end in 2, and so on. But the researchers found that 64 percent of systolic blood pressure readings (the pressure when the heart contracts) and 59 percent of diastolic readings (the pressure when the heart is distended between beats) ended in the value 0. Not only that, but among the non-zero values they found that recorded blood pressures ending in even numbers were significantly more common than those ending in odd numbers, and that five was the most common final digit among those ending in odd numbers. Real blood pressure just does not have this strange grouping around certain numbers! The recorded values are a result of people's tendency to round values to convenient numbers.

Does this matter? The British Hypertension Guidelines give blood pressure thresholds above which they recommend drug treatment.[2] One of these is a systolic blood pressure of 140 mm or higher. But the preferential rounding to values ending in zero (e.g., rounding 137 to 140) means that a significant proportion of patients with this recorded value in fact have true systolic values below 140.

Clearly, in this example, the rounding is a consequence of the nature of the measuring instrument. If the readings were taken from a graduated physical scale, like the rulers used in school, there would be a natural tendency to round them to a convenient value. However, if the values had been given digitally by an electronic readout, they might well have been recorded to more decimal places (still a finite number but a more precise level of rounding). One implication of this is that the continued trend toward automatic and more accurate digital readouts in modern instrumentation is a good thing, at least from the dark data perspective.

That example gives a clue as to when we should expect this phenomenon to manifest itself: we should be especially on the lookout for it whenever humans are involved in reading a value from a physical measuring instrument like a ruler or protractor or a dial. But it can also occur with counts. John Roberts Jr and Devon Brewer asked drug users how many partners they had shared drugs with during the previous six months.[3] While only 2 claimed to have had 9 partners, and only 4 claimed to have had 11 partners, a whopping 39 claimed to have had 10 partners. Likewise, while 21 claimed to have had 20 partners, none claimed to have had either 19 or 21 partners. This all looks very suspicious. It would be extraordinary if people naturally gravitated toward such nicely rounded numbers of partners, and it would be equally extraordinary if the spikes just happened to occur by chance in the sample of people studied. A vastly more probable explanation is that the respondents gave *rough* answers, rounding to the nearest 10.

I've called this phenomenon rounding, but when it is a result of the human stage of the data collection process it also goes under other names, including heaping, piling, peaking, discretizing, and digit preference.

It can also occur in the form of deliberate maximum and minimum limits to values which might be observed. For example, salary surveys are sometimes constructed with a top category of "$100,000 or more" to encourage people to answer who might otherwise be deterred from taking part in the survey at all. This strategy is called *topcoding,* with a corresponding *bottomcoding* if the lower values are truncated.

Ignoring such truncation can lead to serious mistakes. Using a simple average of salaries could obviously be very misleading if that meant you failed to take into account the fact that "$100,000 or more" could mean *a great deal more,* perhaps tens of millions more. Moreover, truncating the largest values in this way and treating them at face value would certainly lead to underestimating the variability of the data.

Summarizing

It is generally not very productive trying to gain insight into a large table of numbers simply by looking at the table. To cope with this sort of challenge we need to summarize the values. Or, put more elaborately, we *analyze* the data, condensing them into summaries and aggregates which are more conveniently grasped by humans. For example, we calculate averages and ranges of values, as well as more sophisticated statistical summaries like correlation coefficients, regression coefficients, and factor loadings. By definition, however, summarization means sacrificing the details: it means darkening the data (*DD-Type 9: Summaries of Data*).

If I tell you that the average weight of American men aged over 20 is 195.7 pounds, then you certainly know something useful.[4] You can compare the value with the average weight in earlier years to see if weight is increasing. You can see how you compare with the average. But what you cannot do is determine the number of men exceeding any particular weight. You cannot tell if this average is produced by relatively few excessively heavy men balancing those below the average, or if there are a great many just above the average. You cannot tell how many men weigh exactly that average to the nearest pound. All these and other questions cannot be answered, because the simple presentation of an average has darkened the data, hiding the individual values.

There are various lessons to take away from this. One is that a single summary statistic, or even any small number of them summarizing the data in different ways (e.g., an average, plus a measure of how widely dispersed the values are, plus a measure of how skewed the distribution of values is) does not tell you *everything* about the data. By darkening the data, summaries might conceal critical information, and you need to be on the alert for this.

A second important lesson is that you need to choose your summary statistic(s) carefully, to answer the question you want to ask. The average value ("arithmetic mean") of incomes of people in a small company, nine of whom each earn $10,000 per year and one who earns $10 million per year, is over $1 million. In many situations that would be misleading information—almost certainly to someone who is applying for a job with that company. For this reason, income and wealth distributions are often characterized in terms of the median value (half earn below and half above this value), rather than the mean. Better still is to give more information about the shape of the distribution—including more summary statistics, such as the number who earn just $10,000 per year, or the maximum value earned.

Human Error

The human rounding we discussed earlier in this chapter wasn't exactly an "error." It was really an approximation, hiding the details, although perhaps in a rather erratic way (not *all* the blood pressure values were rounded to end in 0, for example). But human error can introduce dark data in a much more serious way.

In 2015, Alex Rosetto and Luke Parkin, both second-year students at Northumbria University in the UK, took part in a study to measure the effect of caffeine on exercise. However, a "data error" meant that instead of receiving three times the caffeine in an average cup of coffee, each received 300 times the dose—about 30 grams. (I put data error in quotation marks to draw attention to the fact that it was not the data making the error—it was a human who wrote down the numbers.) To put this into context, people have died after taking 18 grams. Not surprisingly, Alex and Luke spent several days in an intensive care unit undergoing dialysis to remove the caffeine from their blood.

The cause of the overdose was a fairly common kind of error: the decimal point was in the wrong place, so that the data misrepresented what was intended.

A fairly common kind of error? Two days after his 19th birthday, Irish teenager Karl Smith was paid €19,636 instead of the €196.36 he was expecting. Unfortunately, he succumbed to temptation and proceeded to spend it, for which he was sent to jail (though it has to be said that the 17 previous convictions didn't help his case). Likewise, North Yorkshire construction worker Steven Burke was supposed to receive £446.60, but instead his bank account swelled by over £40,000 due to a misplaced decimal point, and he also found himself unable to resist spending £28,000 of it—and received a suspended jail sentence. (There seems to be a common theme here and a lesson to be learned: if you should find your bank account unexpectedly inflated, *don't spend that money!*)

In December 2013, Amsterdam City Council paid its usual housing benefits to nearly 10,000 recipients. However, instead of the amount being in cents, it was wrongly paid in euros, shifting the decimal point two places to the right. This error cost the city €188 million. In 2005, a Lehman Brothers trader mistakenly undertook a trade for $300 million, instead of $3 million. A report on drug prices in the *Times (London)* of 26 May 2018 mentioned the pharmacy in Shropshire, UK, that was paid £6,030 for a medicine which should have cost £60.30, and another in

Greenwich, UK, which was paid £7,450 for painkillers costing £74.50.[5]

As an example of a mistake in the opposite direction, in 2006 Alitalia Airlines intended to offer business class flights from Toronto to Cyprus for $3,900.00, but careless positioning of that pesky dot caused them to be offered at $39.00 per seat, leading to a $7.2 million loss—another casualty of that slippery decimal point.

Presumably, all the foregoing were all just careless slips. I certainly hope so, but it's hardly encouraging to learn that Lord Randolph Churchill, father of the famous British wartime prime minister Winston Churchill, on being shown a column of figures containing decimal points is said to have commented that he "never could make out what those damn dots meant." Disturbingly, at the time he was Britain's chancellor of the exchequer, the equivalent of the USA's treasury secretary.

Basic errors like misplacing the decimal point are a type of data-entry error sometimes called "fat finger" errors. And there are countless other examples, like the incident in 2005 when Mizuho Securities Co. lost over $300 million after offering 610,000 J-com shares for 1 yen each, instead of selling single shares for 610,000 yen each. Or the incident in April 2018 when around two thousand employees of Samsung Securities Co. were supposed to receive a dividend of about $0.93 per share, which totals about 2 billion in *won*, the South Korean currency. Unfortunately, instead they were issued 2 billion *shares*, or more than 30 times the total number of shares in the company, and which were worth about $105 billion.

When these sorts of mistakes occur they are rectified as soon as possible—but often not quickly enough. In the Samsung case, it took 37 minutes to sort out the error, by which time 16 Samsung Securities employees had grasped the opportunity and sold

5 million shares of their windfall. Samsung Securities stock price dropped by almost 12 percent, and at the time of writing it is still down 10 percent, so that the company's value is down some $300 million.

And if you think the $105 billion mistake was bad enough, consider the near-disastrous error on the Tokyo Stock Exchange in 2014. Instead of trading Toyota shares to the value of 1.96 billion yen, a broker accidentally entered this figure in the field indicating the number of shares—equaling a value of $617 billion. An easy enough mistake to make? I've certainly put entries in incorrect boxes on forms before but, thankfully, not with the same potential consequences. Fortunately, in this case, the order was canceled before it could be completed.

Another kind of human error is digit transposition, in which two digits are entered in the wrong order (e.g., 89 instead of 98), or a digit is entered by mistake for another (e.g., 7 instead of 2), or values are repeated (e.g., 222 because of depressing the key for too long), and so on.

Errors like these are simple slips, and, regrettably, humans make mistakes in an unlimited number of ways. For example, units of measurement are sometimes confused—as in the 1998 Mars Climate Orbiter case, when the spacecraft disintegrated on passing too close to Mars because Imperial units of force had not been converted into SI units; or as in the case of Air Canada Flight 143 in 1983, which crashed after its fuel load was measured in pounds instead of kilograms.

A different kind of human error affected the NASA *Genesis* sample-return probe. It successfully captured samples of the solar wind from beyond the orbit of the moon and successfully delivered them back to Earth—until, that is, the very last stage, when it crashed on landing in Utah. The reason: faulty data from the probe's accelerometers, which had been installed backward, so that the probe was accelerating toward the ground instead of decelerating.

A more subtle problem is that the usefulness of data can decrease over time. This is not because data physically deteriorate, like fruit rotting, but because the world around us changes. You might think your savings account is paying 3 percent interest, but if you haven't been paying attention, you might be shocked when you discover that it's less than that and that the rates have been changed without your noticing. Data referring to humans are particularly susceptible to going out of date (*DD-Type 7: Changes with Time*), simply because humans change.

Worse still, and as we shall explore in depth in later chapters, data can even be deliberately distorted: a 1986 study by the U.S. Census Bureau estimated that between 3 and 5 percent of census enumerators engaged in some form of data fabrication because they were too lazy to go to the trouble of actually collecting the data—the curbstoning mentioned in chapter 2.[6] American statistician William Kruskal wrote that "a reasonably perceptive person, with some common sense and a head for figures, can sit down with almost any structured and substantial data set or statistical compilation and find strange-looking numbers in less than an hour."[7] Media research analyst Tony Twyman coined a law now known as *Twyman's law*, which says that any figure that looks interesting or different is usually wrong.[8] Moreover, the sheer volume of numbers recorded every day means we should expect some to be misrecorded. For example, there were nearly 35 billion financial transactions made per day in 2014, a number which has increased since. As I explained in my book *The Improbability Principle*, with so many numbers we should expect to find many that have been misrecorded.

In a related vein, researchers in data mining (the discipline concerned with finding interesting or valuable anomalies in large data sets) say the causes of unusual structures in large data sets are the following, in decreasing order of importance: (i) there's a problem with the data (maybe they have been corrupted or distorted during collection, or some are missing); (ii) the structures are due to chance fluctuations (they are really just rare values which have happened to occur and don't mean anything important); (iii) they turn out to be known about beforehand (like the discovery that people often buy cheese and crackers together); and (iv) they are uninteresting (like the discovery that about half the married people in the UK are female). It's not until all these factors have been accounted for that an unusual structure is real, interesting, and possibly valuable. For us the important point is that most apparent discoveries are illusions caused by glitches in the data.

Given the foregoing, perhaps it's hardly surprising that IBM once estimated that "poor data quality costs the U.S. economy around $3.1 trillion per year."[9] But is this estimate correct?

First, this figure depends on how you define bad data: does it include the cost of rectifying the errors, of checking to see whether there are problems with the data, of the mistakes you make because the data are poor? Second, you might consider that figure in the context of the U.S. national GDP, which is about $20 trillion. Relative to that, the $3.1 trillion seems awfully large—and I can't help wondering whether this estimate itself represents bad data.

Instrument Limitations

Human errors are common enough, but people are not solely responsible for errors: breakdown of physical measuring instruments can also lead to unsuspected dark data problems. At the very least, if an instrument breakdown is not immediately detected, it might give output values fixed at zero or some other number. Think of those television medical dramas in which the heart monitor signal suddenly flatlines. You'd get exactly the same signal if the sensors were pulled off the patient.

One of my postgraduate students worked on a project investigating the effect of adverse weather conditions like high winds and heavy rain on telecommunications networks. His data came from details of network faults and repairs and recordings from weather stations (in fact, linking data sets, as described in the next section). Being a wise student, before he undertook any analysis he studied the data, plotting them this way and that, looking for anomalous values. And he spotted something very strange. On many nights the raw numbers appeared to show dramatic gales occurring at midnight. This was most odd, since no one could remember such storms. Indeed, records from the Meteorological Office showed no such storms. Something funny was going on.

Digging deeper, my student found that at midnight on some nights the anemometer measuring wind speed automatically reset itself. And when it did this it sometimes generated huge apparent spikes in speed. But these were totally artificial. Had my student not understood the need to check the data beforehand, any analysis he might have undertaken would have produced nonsense. As it was, he spotted the problem and was able to correct for it.

Instrument malfunction can lead to expensive consequences. In 2008 a USAF B-2 Spirit stealth bomber crashed in Guam because of incorrect data transmitted by sensors which had become wet. The aircrew believed the plane was traveling at the takeoff speed of 140 knots, when in fact it was flying 10 knots slower.

In the previous section we looked at concealing data by deliberately curtailing values to lie above some minimum and below some maximum. Often, however, this sort of effect arises from the very nature of measuring instruments themselves.

Bathroom scales, for example, will have a maximum value which they can record. Persons heavier than that maximum will know their weight exceeds whatever that maximum is, but their precise weight will remain hidden as dark data. Although this situation is similar to the topcoding strategy discussed earlier, it's not a deliberate choice of those taking the measurement and is called a *ceiling effect*. Similarly, in other contexts there's an analogous lower limit: the value of measurements below some threshold will not be recorded, but only the fact that they are less than or equal to the smallest value the measuring instrument can record. This is like bottomcoding and for obvious reasons is called a *floor effect*. For example, temperatures below the freezing point of mercury cannot be recorded with a mercury thermometer. Ceiling and floor effects lead to *DD-Type 1: Data We Know Are Missing* because we know there is a value, but we don't know what it is—only that it is above or below some limit. This means they also illustrate *DD-Type 10: Measurement Error and Uncertainty*.

Ceiling and floor effects can manifest themselves in rather unexpected ways. For example, there are estimated to be some 10^{24} stars in the universe—that's a 1 followed by 24 zeros. And yet only about 5,000 stars are visible to the naked eye (and because the Earth blocks half the sky, we can see only about half that number from any one point). This means that most data relating to astronomical objects were dark before the telescope was invented: the brightness of these objects lay below the floor of sensitivity of the human eye. Thus, any conclusions about the nature of the universe based on analyzing the few thousand visible stars could easily be very misleading.

Around 1609 Galileo began exploring the heavens with his telescope, with a magnification of about 30, and this revealed the existence of stars which had not previously been suspected. Since then, further advances in technology have continued to reveal more information about the universe. Nevertheless, the fundamental challenge remains that more distant objects appear dimmer, and dimmer astronomical objects are less likely to be seen. Failure to correct for this situation leads to so-called Malmquist bias, after the Swedish astronomer Gunnar Malmquist, who explored it in the 1920s, and it has various subtle implications. For example, although the brightness limits at which both stars and galaxies can be detected are the same, stars are more likely to exceed the detection threshold and be seen, because they form a more concentrated light source. In general, ignoring this differential dark data effect would lead to misunderstanding the structure of the universe.

Telescopes are an example of an advance in technology which reveals the existence of worlds (literally, in the case of telescopes) that we had not previously imagined: they shine metaphorical light into the shadows concealing data. Other instruments serve the same purpose in other domains. Microscopes and medical scanning technologies reveal previously concealed things about the human body, aerial photographs of Earth can tell us about ancient walls and buildings, and seismological and magnetic-field-detecting instruments tell us about the inside of the Earth. All these and countless other instruments extend the human senses, revealing data which were previously dark.

Linking Data Sets

Individual data sets have immense potential for benefiting humanity, but a particular synergy can result when we look beyond data acquired from a single source and consider linking, merging, or fusing data sets. The records in one data set might complement those in another, providing different types of information; they might supplement the other, for example, enabling questions to be answered that either data set alone could not answer; or they might be able to improve accuracy through triangulation and imputation, in which values missing in one data set can be filled in using information from the other.

One obvious group making use of these ideas are forensic statisticians and law enforcement agencies concerned with detecting fraud, but in fact they are applied much more widely. The projects undertaken by the UK's Administrative Data Research Network illustrated the power of such data linkage.[10] This consortium of universities and the National Statistical Institutes of the four countries of the UK aimed to facilitate the linkage and analysis of administrative data for purposes of social science and public policy research. One project brought together data from several sources to explore the impact of housing assistance on homeless people's health conditions and use of healthcare services. Another combined databases to look at the effect of fuel poverty on health. And yet another merged data to look at the relationship between the density of alcohol outlets and population health.

An example of the power of this sort of strategy was a project which linked data from six social services agencies in the United States to gain a clear picture of homelessness in Los Angeles County and drew up a $2 billion program to build 10,000 houses for homeless people with mental health problems.[11]

The potential is unlimited and demonstrates the power of modern data technology to do good. But linking and merging data sets does not come without challenges—and there are dark data risks. To merge data sets there must be common identifiers or sets of identifiers so that a record in one set can be matched with a record in another. Frequently, however, these are not recorded in the same format or style, so that mismatches occur.

Almost always there are records in one database referring to people not included in the other. Duplicate records complicate things further. Methods for matching and linking data so as to minimize the extent of dark data have become a major research area, one which is likely to grow in importance as more and more large data sets accumulate.

So where does all this leave us? Chapter 2 explored different kinds of data, and chapter 3 and this chapter have looked at the dark data risks arising during data collection. These included ambiguous definitions, missing variables, random aspects of measurement processes, instrumental limitations, data coarsening, fat finger errors, and other risks. But there are more: the next chapter looks at an entirely different class of ways in which dark data can be generated.

Chapter 5

STRATEGIC DARK DATA

Gaming, Feedback, and Information Asymmetry

Gaming

The European Union's so-called Gender Directive* rules prohibit insurance companies from using sex when deciding on premiums; that is, they require that sex be treated as dark data when they make such decisions.[1] In principle, this means that, all else being equal, a man and a woman would pay the same premium. In Canada, however, things are different. In 1992 the Canadian Supreme Court permitted sex to continue to be included in risk assessment models. This ruling led one man from Alberta, taken aback by the amount he would be charged to insure his new Chevrolet Cruze, to obtain a new birth certificate stating that he was female. He declared "I'm a man 100 percent [but] legally I'm a woman." Legitimately hiding his true sex in this way saved him $1,100 dollars per year.

Fraud, to be discussed in the next chapter, is an overt attempt to deceive—making people believe things are other than they

*A "directive" requires EU member states to achieve a result without specifying how they must do it. It is contrasted with a "regulation," which immediately becomes enforceable law in all member states simultaneously.

really are by concealing something. In contrast, gaming, as in "gaming the system" or "playing the system," seeks to take advantage of misleading, ambiguous, or unintended aspects of the way things are. The dark data in gaming arise not from deliberate concealment but, rather, as some accidental aspect of the way the system is constructed and which can be taken advantage of. This means that gaming will typically not be illegal: the aim is to remain within the rules while manipulating them to gain an advantage. Gaming is dark data *DD-Type 11: Feedback and Gaming*.

There's a very deep and profound theorem in mathematics named after its discoverer Kurt Gödel, which says—in a vastly oversimplified formulation—that any sufficiently rich axiom system has statements which cannot be proved or disproved within that system. But at a more human level, examples arise in elaborate systems of regulations, which often inevitably have gaps. One such domain in which we regularly see such gaps is the tax regime. Legal tax avoidance schemes spring up to take advantage of the ambiguities or oversights in tax laws. The details will obviously vary both between tax codes and over time as the laws change, but the following are some examples of such schemes which have been perpetrated in the UK:

- avoiding inheritance tax by borrowing against a taxable asset (e.g., your house) and investing the loan in a nontaxable property, such as woodland or a farm;
- buying property through an offshore company, since non-UK residents and companies don't pay UK taxes;
- shifting a company's corporate headquarters to a country with a low corporate tax rate, perhaps by merging with or acquiring a firm in the second country, since there is no worldwide tax authority.

When gaps in a tax system are discovered and people start making extensive use of them, they will be closed, but often that merely leads to even more elaborate structures, which in turn will have their own gaps.

The so-called principal agent problem is another illustration of dark data closely allied to gaming. This problem arises when one person (the "agent") is enabled to make decisions on behalf of another (the "principal"). Obviously, this is a very common situation, as when employees make decisions on behalf of their employer, and politicians act on behalf of their constituents. It becomes a problem when the agent is motivated to make choices which are in his or her own interests but not in the principal's interests. Employees might start to take advantage of their extra knowledge to make choices which benefit themselves rather than their employer, and the politician might begin to act in his or her own interests. This latter is a slippery road which can lead to dictatorship.

Gaming also arises in so-called regulatory arbitrage, which describes what happens when several regulatory regimes apply in a particular situation, allowing organizations (e.g., financial bodies) to choose which regulator applies to them (e.g., by shifting the head office to another country). Clearly, they will choose whichever is most beneficial and might even occasionally reclassify their activities so as to switch between regulators.

Campbell's law is a nice summary of the danger of gaming in public policy contexts. It says: "The more any quantitative social indicator is used for social decision-making, the more subject it will be to corruption pressures and the more apt it will be to distort and corrupt the social processes it is intended to monitor." *Goodhart's law* says something similar, albeit in a milder form: "When a measure becomes a target, it ceases to be a good measure."

Take school grades for example, an indicator of academic performance used to make many decisions in society. Examination of grades over time shows very clearly that the average grades awarded to students tend to drift up over time—so-called grade inflation. A 2018 study of U.S. schools by Michael Hurwitz and Jason Lee found that the number of A averages achieved by those taking the SAT examinations in America rose from 39 percent to 47 percent over the past 20 years.[2] The website GradeInflation.com goes into great detail on grade inflation at U.S. colleges and universities. For example, it shows that grade point average increased from about 2.83 to 3.15 between 1983 and 2013 and did so in a remarkably regular year-on-year way. Such trends have several possible explanations: people are getting cleverer, they are becoming more skilled at answering the sort of questions asked in these tests; or somehow the system is becoming corrupted, so that people tend to get higher scores.

This trend is mirrored in the UK higher education sector, although things are slightly complicated by the vast increase in the numbers of students over time: the percentage of those in the age range 25 to 29 with degrees increased from 13 percent in 1993 to 41 percent in 2015,[3] and by 2017 there were 2.3 million students studying at British universities.

This means that to determine whether standards have changed, first, we should look at proportions rather than absolute numbers of students achieving the various classes of degrees.* Second, we would expect a decrease in the proportions achieving higher degree classes. This follows from the presumption that the universities previously made efforts to select those few students who were the most able and who could benefit most

*The British university system awards degree classes of 1st, 2i, 2ii, and 3rd, in decreasing order of excellence.

from a university education, so that taking a higher proportion of each age group necessarily means taking less able students or those less suited to this type of education and who are consequently less likely to obtain the top degrees. But what do the figures show? In his highly illuminating book *A University Education*, former UK universities minister David Willetts commented that in the year 2000 some 55 percent of students graduated with a 1st or 2i class degree, but by 2015 that number had risen to 74 percent.[4] Not only was that an increase, contrary to what we would expect, but it was also a surprisingly large increase.

What drives such inflation?

University income depends on the number of students admitted, so the more applicants the better. And students will be encouraged to apply to a particular university if they believe studying there will enhance their chances in the job market—which will happen if the students obtain high grades. Since universities in the UK award their own degrees and grade them themselves, this means there is a natural pressure to award high grades. It is the competition between bodies which are able to grade themselves that drives the inflation. The situation would be different if universities used a common set of examinations, and students were graded by a single body: with the present system these standards are dark data. The phenomenon is aggravated by league tables, which rank universities in terms including the numbers of different classes of degree passes, which enables new applicants to preferentially apply to those they believe will give them higher grades.

In fairness, I should add that my description is something of an oversimplification, and there are moderating influences. For example, there is an "external examiner" system by which the quality of the teaching and the level of the degrees is in principle overseen by people from other universities. Moreover, a university which consistently awarded only first-class honors might rank high in the league tables for a while—until it became common knowledge that many students with a "good" degree from this university actually knew very little. Then employers would look elsewhere, and the university's student intake would slump as people learned that its graduates were unemployable.

Schools in the UK are a bit different. Toward the end of secondary school level nationwide public examinations determine who will advance to further school and university education. However, there are several competing boards, each setting their own nationwide examinations. The more students who sit a board's examination, the more money that board makes. And the more high grades a school gets, the better it looks in the league tables. This again may stimulate a race to the bottom—or the top in terms of levels of mark awarded—although some have argued that there is no evidence that the boards have different levels of difficulty.

In addition, schools can have a say in who is admitted to them. Indeed, once students are admitted, the schools can decide which ones to allow to take certain public examinations. Clearly, by permitting only the most able students to sit the examinations, schools can give a distorted impression of their effectiveness, in a clear case of *DD-Type 2: Data We Don't Know Are Missing*. If the performance of an organization is measured by its success rate, then choosing conditions in which it is most likely to succeed can make it look good. Writing in the *Times* (London) in August 2018, Rachel Sylvester said: "A growing number of schools are gaming the exam system to boost their league table rankings to the detriment of pupils. . . . Private schools routinely encourage children to drop subjects in which they are not predicted to get top marks in order to keep up their overall grade average."[5] Pupils expected to perform poorly might even be asked to leave a school, so that its apparent performance looks better. Sylvester cites the UK schools inspectorate Ofsted (Office for Standards in Education, Children's Services and Skills) as finding that 19,000 pupils were removed from the school roll just before taking their GCSEs (a national examination taken at age 16). The potential impact of such an action is obvious, both on indicators of the effectiveness of schools and also on individual pupils.

Just about any domain in which performance is evaluated can provide examples of such gaming. The following are a few from very different domains:

- Surgeons can achieve a high success rate by avoiding the more serious cases, which naturally have a more doubtful outcome. More generally, even without overt choice of whom to operate on, patient populations are likely to differ between surgeons. This means that even surgeons of the same ability should be expected to have different success rates.
- Emergency service response times can be manipulated by redefining the precise nature of the emergency. The *Telegraph* (London) of 28 February 2003 reported that "according to [the Commission for Health Improvement], West Yorkshire Metropolitan Ambulance Service NHS Trust (WYMAS) downgraded some category A calls if the ambulance crew arrived at the scene and decided the call was not, in hindsight, serious enough to warrant a category A response. . . . The watchdog also found a substantial time lag between the time a call was received and the time the trust started the clock to time its response."[6]

- As mentioned in chapter 3, the definition of unemployment can be changed to adjust the figures: should gig workers or part-timers who are seeking full-time employment be considered unemployed? An extreme of such manipulation is illustrated by the difference between the unemployment estimate for the end of February 2017 provided by the U.S. Bureau of Labor Statistics, which was 4.7 percent, and that provided by President Donald Trump, which was 42 percent.[7] The latter figure was obtained by including as unemployed everyone above 16 who was not in the labor force—stay-at-home parents, full-time students, retired grandparents, and so on. That's not the definition normally used by economists. In general, in such cases it is not a matter of one definition being "right" and the other "wrong." They are simply different (*DD-Type 8: Definitions of Data*)—and may well be more or less useful for specific purposes.
- Police forces can appear to do better by reclassifying offenses as less serious. For example, according to Glasgow's *Herald* of February 2014, "officers feel the need to massage the figures by downgrading crimes. The classification system has two tiers: 'crimes,' which were down 13 percent last year; and 'offenses,' which were slightly up. In 2012–13, there were 273,053 recorded crimes, but there were nearly twice as many offenses."[8]

And you may well know that websites can be manipulated so that they appear earlier in searches—thus helping company sales or apparent relevance of blogs, for instance.

All these are examples of making choices and manipulating definitions so as to conceal something or to present it in a different light. They hide something which would adversely impact an organization were it to become known, or they make apparent something which could benefit someone or some organization but which might otherwise have been unnoticed.

Feedback

Good scores on a test can encourage more effort, which might lead to more success, which in turn might lead to even greater effort. The final data are real enough, but they are not what they would have been had they not been measured. Although not completely dark, they are hiding what was there before we intervened. I say "intervened," but our intervention was not a deliberate attempt to change the figures. It was an attempt to discover what the values were. It is the very making of this attempt which changed the figures, so that we are left with values different from those we originally set out to measure.

The preceding is an example of a feedback mechanism, in which data we have measured loop back to influence and change the values themselves. Such mechanisms are ubiquitous. They occur in physical systems: you might be familiar with them in sound systems, as when a microphone picks up the output of a loudspeaker it is linked to, feeding it back to the loudspeaker, where it is picked up again by the microphone, and so on in a loop, growing louder each time and resulting in a howling sound. They occur in biological systems: in blood clotting for example, damaged tissue releases a substance that activates blood platelets, and these in turn release more of the same substance, leading to the activation of more platelets. And such mechanisms occur in psychology: knowing that you are being observed in a task can stimulate you to try harder (the Hawthorne effect mentioned in chapter 2). They are particularly striking when they occur in the world of finance in the form of "bubbles."

Bubbles in financial markets are sudden dramatic rises in stock prices (or some other asset) followed by an equally dramatic drop. The changes in prices do not reflect any real underlying changes in value but, rather, arise from greed or lack of critical assessment of the underlying value of the asset: the mistaken belief that the underlying value has really increased. The fundamental point is that while underlying value is one of the factors influencing share prices, the bottom line is what others are willing to pay. Thus we have the analogy with a beauty contest made by the eminent economist John Maynard Keynes that "it is not a case of choosing those [faces] which, to the best of one's judgment, are really the prettiest, nor even those which average opinion genuinely thinks the prettiest. We have reached the third degree where we devote our intelligences to anticipating what average opinion expects the average opinion to be. And there are some, I believe, who practice the fourth, fifth and higher degrees."[9]

History is littered with examples of financial bubbles.

A major one occurred in the early part of the eighteenth century, when the French tried to introduce paper banknotes—money had previously been based on physical precious metals. The issue of the banknotes was associated with a financial bubble so dramatic that it destroyed the French economy and delayed the further introduction of paper notes for 80 years.

It all began in 1716 when Scottish economist John Law persuaded the French government to allow him to set up a new bank, the *Banque Générale*, to issue paper banknotes supported by the bank's reserves of gold and silver. That arrangement would probably have been fine, but Law had much grander plans. The next year, in 1717, he went on to persuade the French government to hand control of trade between France and its Louisiana/Canada colony to him. The colony was vast, stretching for 3,000 miles from the mouth of the Mississippi River through Arkansas, Missouri, Illinois, Iowa, Wisconsin, and Minnesota and including parts of Canada. To finance the company, Law sold shares for cash and state bonds. Since the colony was reputed to be rich in gold and silver, the scheme attracted willing purchasers. But that was not enough for Law. He went on to acquire a monopoly over the tobacco trade between France and Africa, and of companies trading with China and the East Indies. From there Law's "Mississippi Company" bought the right to mint French coins and also the right to collect most French taxes. All these activities were funded by issuing additional shares in the company.

The growth in the Mississippi Company was matched by the rise in value of the shares, increasing by a factor of 20 over the year 1719. And the rapidly rising shares attracted purchasers—even to the extent that soldiers sometimes had to be called out to control the milling mobs of those who wanted to buy shares. And, as always with speculative bubbles, people who can ill afford to lose the money started to invest.

It is a characteristic of bubbles, whether they are real or metaphorical, that they burst.

The turning point for the Mississippi Company came in January 1720, when some investors started to sell shares to cash in on their gains. In general, in such cases, initially a few people sell. But these sales mean prices stop rising so sharply or even start to fall. That in turn leads others to sell, believing that the peak has been reached and hoping to cash in before the price falls too much. And that in its turn leads to yet more selling. And suddenly prices are crashing, typically at a faster rate than they rose.

Law made valiant efforts to recover, limiting the size of payments in gold and devaluing shares in the company, among other ways. But by December 1720 the price of shares had dropped to one-tenth of their peak value. Law became the subject of mob vilification and in the end left France, dying poor in Venice.

The case of John Law and the Mississippi Company is extraordinary enough, but perhaps an even more famous historical example is the Dutch tulip market bubble.

At the end of the 16th century tulips were introduced into Holland from Turkey. As a new kind of flower, they were already pricey, but when tulips with unusual variegated colors (actually caused by a plant disease) were developed, they became even more so. Limitations on the supply of tulip bulbs stimulated competition, and prices for future purchases began to rise. The rising market stimulated others to buy, believing they would be able to sell the bulbs later. And then the race was on. Savings, houses, and estates were sold to yield money to buy tulip bulbs. Clearly, these hugely inflated prices did not reflect any underlying real value, with the inevitable consequence once people began to cash in on the increased value of their purchases. A catastrophic crash in the price of tulip bulbs followed. People lost their fortunes, even their homes.

The story is a familiar one of boom and bust, so you might be forgiven for thinking that only the naïve would fall for it. But things look different from the inside, as illustrated by the story of Isaac Newton and the South Sea bubble. In a contemporaneous development paralleling John Law's Mississippi Company, the English government granted the South Sea Company a monopoly on trades in the South Seas. Appreciating the advantages of a monopoly, investors began to pour in. Share prices rose dramatically. Isaac Newton bought some shares and then sold them in early 1720, making a tidy profit. Unfortunately, as the share prices continued to rise he decided he had sold too early, and he jumped back in. The shares continued to rise, then peaked, and crashed later in 1720. And Newton lost almost all his life savings. If this can happen to Isaac Newton, then it can happen to anyone.

Those are historical examples, but the sound of bursting financial bubbles is not something we hear merely as echoes from history.

The so-called dot-com bubble was triggered by a wave of interest in hi-tech companies stimulated by the development of the internet. As soon as shares of the startups were floated, many of these new companies were valued in billions of dollars. One consequence was that shares in the NASDAQ Composite index, representing the exchange where these companies were traded, also shot up in price, 10-fold between 1990 and 2000 (not quite on the Mississippi Company level, but startling nonetheless). Then, the crash occurred as people began to realize that the stocks were overvalued—that their price was illusory and did not properly reflect underlying reality in terms of economic value. By October 2002 the NASDAQ index had fallen to just over one-fifth of its previous value. As with the Mississippi case, the knock-on effects were substantial—an economic recession in the United States resulted.

The NASDAQ bubble was closely followed by the U.S. housing bubble—indeed, it has been suggested that the need for investors to find homes for their funds after selling NASDAQ shares led to a shift toward real estate. House prices began to increase dramatically. Subprime lending and other activities indicative of a bubble occurred. At least, that is, until 2006 when the peak arrived, and things went horribly awry. Within three years the average house price had dropped by one-third. This crash in turn contributed to a global recession, the largest since the 1930s.

The following is a final example of data distorted by feedback. This time the feedback explicitly served to conceal some of the data.

In 2011 England and Wales launched an online system of crime maps, enabling users to see what crimes had taken place in the vicinity of any location. The then British home secretary (and later prime minister) Theresa May said: "I think people are going to welcome the fact they can really see what's happening with crime in their area, not just on their street but in their neighbourhood." In 2013 the New York Police Department released a similar interactive map, and nowadays such systems are commonplace. The obvious merit of them is that they allow people to make informed decisions—about where to buy or rent, about whether to walk down a particular street at night, and so on. Of course, as with any large databases, they are not perfect, and sometimes errors slip through: "The crime maps show Surrey Street in Portsmouth, Hampshire, [UK], as having 136 crimes, including burglary, violence and anti-social behaviour in December. . . . But the street, which is less than 100 m long, is only home to a pub, a car park and a block of flats."[10] Either it's a street to be avoided at all costs, or there's a problem with the data.

But, data errors aside, there are more subtle issues with the crime-maps idea, issues involving dark data and feedback. Attention was drawn to these when the UK Direct Line Insurance Group carried out a survey and reported that "10% of all UK adults would definitely or probably consider not reporting a crime to the police because it would show up on an online crime map, which might negatively impact their ability to rent/sell or reduce the value of their property."[11] Instead of showing where incidents had occurred, the maps were at risk of showing where people did not mind reporting them. This is not at all the same thing, and anyone basing decisions on such data could easily be misled.

Finally on the subject of feedback, one of the key psychological drivers behind bubbles is confirmation bias, which we have already encountered. This leads us to subconsciously seek information which supports our point of view and to tend to ignore data not supporting it. In the world of finance, as elsewhere, people are naturally pleased to see information that provides support for the decisions they want to make or perhaps have already made.

The term *echo chamber* has been appropriated from the world of acoustics to describe situations in which beliefs, attitudes, and opinions are fed back in loops, reinforcing themselves. In the context of social networks, this feedback can exaggerate marginal beliefs, leading to polarization and extremism. The principle is straightforward. Someone makes an assertion which is picked up and repeated by others and eventually makes its way back to the originator. This person, unaware that it is his or her own assertion being fed back, thinks "See, I knew it! Others think so, too!"

This kind of process is one of the powerful engines underlying the propagation of false facts, fake news, and absurd conspiracy theories. Often, the cycle is accidental, as rumors swirl faster and faster, but individuals have been known to use such a mechanism to deliberately disseminate untrue information. Likewise, governments have been known to spread such misinformation to destabilize other regimes or to sow confusion to prevent their acting in a coordinated way. Creating deliberately incorrect and misleading information in this way is clearly potentially even more damaging than merely hiding the truth and concealing it as dark data.

Information Asymmetry

Information asymmetry is a general term for a situation in which one party has more information than another (*DD-Type 12: Information Asymmetry*). Put the other way round, for one of the parties some of the data are dark. Clearly, this places that party at a disadvantage in negotiations or conflicts. Let us look at some examples.

In a 1970 paper with the wonderful title "The Market for 'Lemons': Quality Uncertainty and the Market Mechanism," Nobel Prize–winning economist George Akerlof presented a beautiful parable describing how information asymmetry could have dire consequences. A "lemon" here is slang for a car that turns out to be of poor quality or defective after it has been purchased. It's the opposite of a "peach," which is high quality and sound.

Buyers of a used car won't know if it is sound or defective at the time they buy it. Other things being equal, a car they buy might equally turn out to be a lemon or a peach. They will therefore be prepared to pay only the average price of a lemon or a peach. But the sellers will have the advantage of knowing—and they won't want to sell a peach for such a price. So they'll hang on to the peaches, selling only the lemons. Buyers will learn they are being sold only lemons and will accordingly offer lower payments, in turn discouraging the sellers even more from selling peaches. A feedback loop will have been established, driving peach holders away from the market and driving down both prices and the quality of sold cars.

In the worst case this can cause the bottom to fall out of the market altogether, so that it ceases to operate.

Asymmetric information is often of the essence in military conflicts. For example, if one side knows more about the disposition of troops than the other side, it can gain an overwhelming advantage. This concept is what lies behind the use of data collection strategies ranging from simply sending scouts ahead to explore the enemy lines to the use of drones, satellite photography, and hacking of telecommunications.

The same applies to spying, in which each side tries to discover data the other side wishes to keep dark, as their release can do a tremendous amount of damage. In 2010, the U.S. army intelligence analyst Bradley/Chelsea Manning leaked a large number of sensitive documents via WikiLeaks, with the result that the lives of political dissidents and others were put at risk.

In some contexts, regulations have been introduced in attempts to tackle asymmetric information. As Arjan Reurink puts it for the world of finance: "To facilitate the provision of information to the market and to ameliorate the problem of information asymmetries, financial regulators have imposed *disclosure requirements* as a central pillar of financial market regulation in all developed financial markets. Disclosure requirements prescribe that issuers of financial instruments and providers of financial services disclose to the market and their counterparties all relevant information, that they do so in a timely manner, and that they make sure that all market participants have equal access to this information."[12] Put another way, these regulations seek to enforce *transparency*, so that data which would otherwise be dark can be seen.

In general, the lesson to learn from this discussion is the importance of being on the lookout for asymmetric information and to ask yourself what might *he or she* know that *you* don't?

Adverse Selection and Algorithms

Rich Caruana and his colleagues described a machine-learning system for predicting the probability that patients who developed pneumonia would die from the illness. It was usually pretty accurate—except, that is, when patients also had a history of asthma.[13] It predicted a lower risk of dying of pneumonia for these than for other patients. This was clearly an unexpected finding and seemed counterintuitive. How can a further complication interfering with breathing make things better? It could indicate a major discovery—some unsuspected biological mechanism that causes asthma to be protective against pneumonia—but that would be surprising. On the other hand, it might mean that unsuspected dark data are leading to us being misled, so that the conclusion cannot be trusted.

In fact, close examination revealed that the machine-learning system *was* wrong, and its decision was indeed a consequence of dark data. It turned out, unsurprisingly, that patients with a history of asthma were actually high risk—so high in fact that they were sent to the intensive care unit, where they received first-class treatment. And the treatment there was so good that it reduced their risk of dying from pneumonia. Then, the system, not being aware that such patients had been treated differently, merely saw that patients with asthma had a reduced risk. So, quite naturally, it instructed doctors to send such "low-risk" patients home.

The fundamental problem here is that the machine-learning algorithm did not see all the relevant data. In this case, it did not know that the asthma patients were treated differently. However, the problem of algorithms seeing distorted data sets is both ubiquitous and pernicious. And sometimes it results from the best of intentions, as we shall now see.

As we saw at the start of this chapter for the particular case of insurance, many countries explicitly legislate against discriminatory or unfair treatment of specific groups. In the UK, for instance, the Equality Act of 2010 is designed "to make provision to require Ministers of the Crown and others when making strategic decisions about the exercise of their functions to have regard to the desirability of reducing socio-economic inequalities . . . to prohibit victimization in certain circumstances;

to require the exercise of certain functions to be with regard to the need to eliminate discrimination and other prohibited conduct . . . to increase equality of opportunity . . ."

The act then goes on to define direct discrimination: "A person (A) discriminates against another (B) if, because of a protected characteristic, A treats B less favourably than A treats or would treat others." It describes specifics for each of several protected characteristics. This means it is prohibited to treat someone less favorably than another on the basis of a group classification—that they are male or belong to a certain race, for example. The act then defines indirect discrimination: "if A applies to B a provision, criterion or practice which is discriminatory in relation to a protected characteristic of B's."

The United States has a similar law, with "disparate treatment" meaning that someone is deliberately treated less favorably than someone else because of a protected characteristic, while "disparate impact" involves practices that appear to treat groups equally but nevertheless have more effect on some groups than others.

Protected characteristics, which vary a little from country to country, include attributes such as age, being or becoming a transsexual person, being married or in a civil partnership, being pregnant or on maternity leave, disability, gender reassignment, race (including color, nationality, ethnic or national origin), religion, belief or lack of religion/belief, sex, and sexual orientation. The law basically says that protected characteristics must be treated as dark data; they must not influence your decisions. So let's look at a couple of areas in which this law has had an impact.

As we have seen, credit scorecards are statistical models yielding scores showing how likely an applicant is to default on a loan (for example). These models are constructed using past data describing a sample of customers and whether or not they defaulted. People with similar characteristics to those who defaulted in the past might be expected to be higher risk. Now it seems obvious that when constructing such scorecards we will want to make them as accurate as possible. For example, we would like to be confident that if the scorecard estimates that 10 percent of applicants with certain characteristic values will default, then about 10 percent will actually default. It could be disastrous for our business operation if 80 percent of applicants like this defaulted.

To make the scorecard as accurate as possible, it would be wise to use all the information we can and not ignore information which might be helpful. And at this point, the reader will possibly have spotted a problem. For accuracy, it would be wise to include the protected characteristics such as those listed earlier, but for sound reasons the law prevents us from doing so. It says we must not include these in our decision-making process.

Obviously there are ways around this restriction. If we cannot include age in the scorecard, perhaps we can include something highly correlated with age as a substitute. But the lawmakers are wise to this way of letting protected characteristics creep in through the back door. A U.S. congressional report on credit scoring said: "Results obtained with the model estimated especially for this study suggest that . . . certain credit characteristics serve, in part, as limited proxies for age." And the report pointed out that "a result of this limited proxying is that the credit scores for older individuals are slightly lower, and those of younger individuals somewhat higher, than would be the case had these credit characteristics not partially proxied for age."[14]

To prevent this surreptitious use of protected characteristics, as well as prohibiting the use of the characteristics themselves, the regulators could also prohibit the use of variables which are correlated with them. Unfortunately, this solution has two problems.

First, as the congressional report went on to say, "Analysis shows that mitigating this effect by dropping these credit characteristics [correlated with age] from the model would come at a cost, as these credit characteristics have strong predictive power over and above their role as age proxies." That means that dropping these correlated characteristics from the scorecard may mean sacrificing legitimate useful information.

Second, the fact is that, at least as far as humans are concerned, most things are correlated at some level. We could easily end up throwing away all the predictive information. We'd end up with a scorecard which classified everyone the same way, either all as "bad risk" or all as "good risk."

There is another, even more important, point. If we were able to omit sex, say, as well as all characteristics correlated with it from the model, then the predictions for men and women would be fair in the sense that a man and a woman who were identical on those characteristics which *were* used would have the same scores. Now, the fact is that, in general, women are less risky than men: other things being equal, they are less likely to default. This means that the consequence of forcing men and women who were identical in terms of the recorded data to have the same scores would unfairly penalize women—it would overestimate their probability of defaulting—while unfairly rewarding men by underestimating their probability of defaulting. And the probability estimates would be reflected in the premiums paid. You might think that this hardly seems fair.

It all comes down to precisely what you mean by "fair."

A U.S. study showed that for men the average credit score was 630 out of 850, whereas for women it was 621. This difference can be at least partly explained in terms of differences between the groups, such as men having higher wages on average, since income was one of the factors included when calculating the score. Commenting on this study, Stew Langille, Credit Sesame's chief strategy officer, said: "In some ways, it's good news that the study shows there isn't a huge discrepancy between the credit scores for men and women, but it doesn't feel as equitable as it should."

Credit scoring is not the only place this form of dark data arises. Insurance has the same sort of structure as credit scoring: the aim is to build a statistical model to predict the probability that an event will occur—dying, falling ill, a car crash, and so on. Unlike credit scoring, insurance predictions in the European Union could, until recently, be based on any data you liked so as to enable you to make the best possible predictions. But, as we saw at the start of this chapter, in 2004 the EU Gender Directive was introduced to counter sex-based discrimination. This directive said that EU insurers must not include sex as a factor in determining premiums and payouts. It said sex was to be treated as dark data, bringing it into line with credit scoring.

However, the EU Gender Directive had an opt out clause. It permitted "proportionate differences in individuals' premiums and benefits where the use of gender is a determining factor in the assessment of risk based on relevant and accurate actuarial and statistical data." So men and women identical in all other characteristics in the statistical model could be charged different insurance premiums, provided data showed they had different risks.

This is all very well, and it adopts one of the perspectives on what is meant by "fair" we noted earlier. But then, in 2008, a lawsuit was brought to the Belgian Constitutional Court claiming that the opt out was incompatible with the principle of equal treatment for men and women. The legal process ground slowly on, and finally, in March 2011, the European Court of Justice decided that the opt out was to be treated as invalid from 21 December 2012. Thus, it would be illegal to have differential insurance premiums based on sex even if data showed that otherwise identical men and women had different risks. From then on, sex had to be treated as dark data.

In the case of auto insurance premiums, for example, females previously paid lower rates, since the data showed they were less likely to have accidents. But after the change in the law, such differences were no longer allowed. The impact is illustrated by a table published in the *Telegraph* (London) of 21 January 2013.[15] The average insurance premium for men (the riskier group) before the ruling was £658. After the ruling it was £619. In contrast, the average premium for women before the ruling was £488, which increased to £529 after the ruling. In the riskiest group, ages 17–18, the premium for men shrank from £2,298 to £2,191, while that for women grew from £1,307 to £1,965.

But there's more. The new premiums mean that the riskier group, the men, will find it easier to buy insurance and therefore will be more likely to take to the road, while the less risky group, the women, will be less likely to do so. This is hardly beneficial to society! Again, the situation hinges on a particular interpretation of "fair."

In general, insurance premiums are based on estimates of the risk that some adverse event will occur, that someone will have a car crash or fall ill, for example, and will need to file an insurance claim. And these risks are based on past data. To take the health example, people can be divided into groups on the basis of their characteristics (age, sex, body mass index, past medical history, and so on), and the data will show what fraction fell ill in each group with identical characteristics. These fractions can then be used as estimates of the probability that an individual person with the same characteristics as a particular group will fall ill in the future. And this probability will be used to determine the premium that each should pay, with all such persons in the group paying the same premium because they are regarded as having the same chance of falling ill. The job of the actuary is to make such calculations.

But now let's look at this group of people as time progresses. Members of the group will change in different ways. Some will gain weight, others will stop smoking, yet others will stop paying their premiums and drop out, and so on. The risk that each member of the group will fall ill will change, and it will change in different ways. Relative to the average risk in the group, some will become less likely to fall ill, while others will become more likely. Some will become less likely to make an insurance claim, while others will become more likely.

Those less likely to fall ill will notice that their healthy condition means they can reduce the premiums they pay by transferring to another insurer. That leaves the riskier people paying the original premium. Data gradually accumulate showing the insurance company that the total premiums paid by the riskier people are unlikely to cover the cost of claims. It therefore increases the premiums. The cycle then repeats—forming what is called an insurance "death spiral" of gradually increasing costs. Remember George Akerlof's market for lemons?

The fundamental point here is that insurance is based on *averages*. Everyone in the initial group was taken as having the same risk, even if they really had somewhat different risks. Treating them all the same essentially treats their deviations from the average as dark data (*DD-Type 9: Summaries of Data*).

Replacing data with averages is an example of obscuring values by summarizing or aggregating, and it's not a hypothetical or theoretical phenomenon. Let's look at the Affordable Care Act, the health plan signed into law in 2010 and often called Obamacare.

A clause in the plan, the so-called individual mandate, required most Americans to buy healthcare insurance or incur a fine, except in certain special circumstances. That meant that healthy people, who had a low risk of falling ill and requiring expensive medical treatment, were included. This in turn meant that on average the pool of insured people had a lower risk—so that premiums could be lower. However, a 2017 Senate bill removed this mandate, so that people would not have to have health insurance (at least—starting in 2019—all these legislative changes are timed to take effect in the future relative to when they were enacted, provided they haven't been amended in the interim!). As a consequence we might expect the lower-risk individuals to drop out disproportionately more than the higher-risk individuals, so that more treatments would be needed on average, incurring more expense. And this in turn would mean higher premiums. In fact, the Congressional Budget Office predicted that the removal of the individual mandate would cause 13 million people to decide not to take out insurance by 2027, resulting in a 10 percent per year increase in premiums. Not everyone agreed with its estimates: Standard & Poor estimated a lower figure, of between 3 and 5 million people not taking out health insurance over the 10-year period. Either way, it does not look very promising.

There are many other complications. One of these is the fact that U.S. *insurers* may choose not to take part in the program. This is another possible source of adverse selection, affecting the data and the entire insurance system. At the time of writing, the situation is in flux, and it will be interesting to see how it pans out.

In this chapter we have seen how ambiguities and oversights in regulations can be taken advantage of, how observed data values themselves might influence the data generation process, how information asymmetry can give one person an advantage over others, and the impact of these aspects of dark data on algorithms. To make things worse, these aspects of dark data can work together, as the death spiral in insurance shows. But manipulating the rules is one thing: *deliberately* fabricating data is something else. And that is what we explore in the next chapter.

Chapter 6

INTENTIONAL DARK DATA

Fraud and Deception

Fraud

Some fraudsters become famous. Victor Lustig had nothing to do with the Eiffel Tower, but this did not stop his selling it. In 1925 he gathered together a group of scrap-metal dealers and told them that the cost of maintaining the tower was so expensive that the city of Paris had decided to sell it for scrap. This did not seem unreasonable, because the tower was originally intended to be only temporary, having been built for the 1889 Paris Exposition. Lustig explained that, naturally, such a sale would cause something of a public outcry, so it was important to keep it secret until the deal was signed off. With forged documents describing him as the deputy director general of the Ministère de Postes et Télégraphes, Lustig took the scrap-metal dealers on a tour of the tower and requested bids. This allowed him to identify one of them, André Poisson, as a suitable target, and Lustig arranged a private meeting with him. At that meeting, Lustig deliberately gave Poisson the impression he could be bribed to accept Poisson's bid for the tower. And then, pocketing both the bribe and the payment, Lustig fled to Austria, later becoming known as "the man who sold the Eiffel Tower."

This true story is a masterpiece of layer upon layer of deception, in which the true state of affairs is hidden (*DD-Type 13: Intentionally Darkened Data*). Yet a further layer was added when Poisson turned out to be too embarrassed to tell people he had fallen for a confidence trick and kept it secret.

Lustig was also famous for his "money printing machine." This was a machine which appeared to print $100 bills. Lustig convinced potential purchasers by showing it in action, very slowly printing out a couple of bills to demonstrate it. Unfortunately, by the time the buyers realized the $100 bills were real bills merely being pushed out of the machine, Lustig (and the $30,000 he charged for the machine) were long gone. And the victims could hardly complain to the police that they had been defrauded in their attempts to buy a counterfeiting machine. Again, layers of deception were involved, each hiding the truth.

As Lustig's con tricks illustrate, the key to frauds is the concealment of information about the real situation: the hiding of data. But such deceptions also often rely on the tendency of the human mind to make snap judgments rather than go to the trouble of painstakingly weighing the evidence and looking carefully at the data. Nobel laureate Daniel Kahneman has explored this tendency in depth, describing it in his best-selling book *Thinking, Fast and Slow*. He distinguishes between what he calls *System 1* and *System 2* thinking. System 1 is quick, instinctive, and changes with emotional state. System 2 thinking is slow, careful, and logical. System 1 allows us to respond quickly to changing events in the world and to make snap judgments which we hope are correct. But such snap judgements can be wrong, and they are subject to various subconscious biases like those mentioned in chapter 2. In contrast, System 2 looks at the evidence, balances the facts, and arrives at a conclusion only after carefully weighing the evidence for and against it. System 2 would take into account the possibility that the data might not be as they appear and that there might be something missing.

The *New Oxford Dictionary of English* formally defines *fraud* as wrongful or criminal deception intended to result in financial or personal gain. While the goal is often to benefit financially, it can be something else—perhaps to gain power, reputation, or sexual favors, or to commit terrorist offenses. And, unfortunately, it can arise in all human activities. As we shall see, it occurs in financial transactions ranging from fraudulent credit card purchases to insider trading in the financial markets. Fraud occurs in counterfeiting, where the aim is to conceal a true nature, so that we are misled into believing an imitation is the real thing—in art, banknotes, pharmaceutical drugs, consumer goods such as handbags and clothes, and many other areas. It occurs over the internet. It occurs in literature, through plagiarism. It occurs in elections, in which the aim is to conceal the true voting outcome so that power can be gained or retained. And as we shall explore in chapter 7, fraud occurs even in science, where the underlying motivation might be reputational enhancement or the sheer overwhelming confidence that *you are right*, even if it is difficult to find genuine data which support that belief.

Véronique Van Vlasselaer and her colleagues give another description of fraud in their outline of a tool for detecting Social Security fraud, characterizing it as "uncommon, well-considered, time-evolving, carefully organized, and imperceptibly concealed crime."[1] The last characteristic, concealment, again highlights the dark data nature of fraud: fraudsters try to cover their tracks, at least for a while. Bart Baesens and his coauthors also comment on this in their book *Fraud Analytics*, saying, "fraudsters tend to try and blend in as much as possible into the surroundings. Such an approach reminds [us] of camouflage techniques as used by the military or by animals such as chameleons and stick insects."[2] So it's not a uniquely human practice. In fact, it appears to be pretty universal in the animal kingdom, ranging from the insects which Baesens mentions, through striped tigers to decorator crabs (which disguise themselves using material from their surroundings, a strategy also used by finches). Some animals adopt a contrary strategy, concealing themselves in plain sight. For example, the ring markings of extremely dangerous coral snakes are mimicked by harmless kingsnakes.

Fraud is now believed to be the most common type of crime. *Crime in England and Wales: Year Ending June 2017* reports: "The latest [crime survey for England and Wales] estimate is 5.8 million incidents of crime in the year ending June 2017. . . . However, these headline estimates do not yet include fraud and computer misuse offences. . . . Including fraud and computer misuse, there were an estimated 10.8 million incidents of crime in the year ending June 2017."[3] There were about the same number of fraud and computer misuse incidents as all the other crimes put together. The cost of (detected) e-commerce fraud was £178 million in 2007, but this amount had shot up to £308 million in 2016. In a 2009 meta-analysis of estimates of the total amount of all kinds of fraud in the UK, my then PhD student Gordon Blunt and I found estimates ranging between £7 billion and £70 billion, depending on precisely how fraud was defined.[4]

Since the web is a worldwide system, it is obviously not only the UK which has experienced an increase in rates of fraud. The U.S. Federal Trade Commission's report *Consumer Sentinel Network Data Book* describes the increase in the number of fraud complaints in America between 2001 and 2014.[5] The trend, shown in Figure 5, speaks for itself.

There are two complementary faces of dark data in fraud. While the fraudster wishes to conceal from you the reality of who

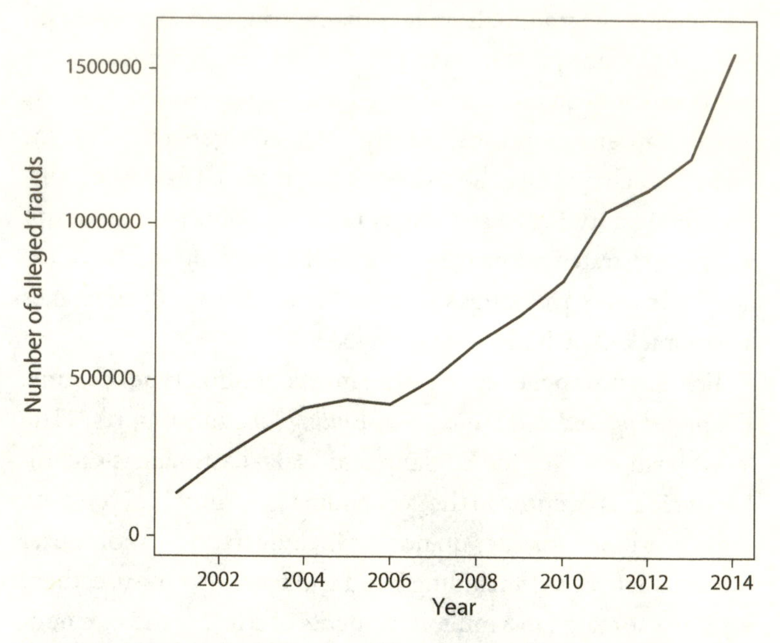

FIGURE 5. Number of consumer complaints about fraud between 2001 and 2014.

he or she is and of the situation facing you (think of Lustig and his scrap-metal merchants), you wish to keep certain data (such as your passwords) secret from others to prevent fraud. This chapter explores the first of these, in just a few of the unlimited number of domains where fraud occurs. We shall look at the second in chapter 9.

Identity Theft and Internet Fraud

There's a famous cartoon by Peter Steiner, published in the *New Yorker* in 1993 and subsequently widely reproduced around the world, featuring two dogs at a computer. One is saying to the other: "On the internet, nobody knows you're a dog." The fact is that the World Wide Web permits the ready concealment of identity, with all the potential for fraud that brings. Furthermore, it can readily facilitate the fraudulent adoption of someone else's identity—identity theft.

The web has made it vastly easier to uncover personal information which can be used to enable crooks to masquerade as someone else, but this kind of deception didn't begin with the web. Indeed, the term *identity theft* was first used in the 1960s. Methods for obtaining confidential personal information used before the advent of the web included simple phone calls (the most common route prior to the development of the internet) and "dumpster diving"—looking through your rubbish to find identifying information on discarded documents like old bills and payslips. And just because the web has arrived does not mean these earlier methods are not still used. Phone scams remain commonplace, often with callers tricking people into calling a bank on a false number and then happily giving out their passwords and other identifying information.

Another earlier tool used in identity theft is the "Day of the Jackal" scam, as used in Frederick Forsyth's bestselling book of that name. This tactic involves obtaining the birth certificate of someone who has died and then using this document to bootstrap the acquisition of other personal documents like passports. While in this case the person whose identity has been stolen clearly does not suffer, the crook can then freely use the identity for other nefarious purposes. Using this scam, 39-year-old Gerald Duffy stole the identity of Andrew Lappin, who had been killed in a road accident in 1972, when he was just three years old. Duffy used this false identity to open a bank account and obtain credit cards.

In a more bizarre example, New Zealand member of parliament and lawyer David Garrett obtained a passport in a dead child's name by first obtaining a birth certificate. He claimed to have stolen the name as a prank after reading about it in Forsyth's book, just to see if it could be done.

Taking the place of someone who has died—or perhaps even whom you killed—seemed to be a familiar strategy for stealing an identity before the advent of the World Wide Web (so perhaps we should be grateful to the web for making that reason for murder unnecessary!).

Child identity theft is particularly problematic, as it may be undetected for a long time, possibly until the child grows up. The *New York Times* reported the case of Gabriel Jimenez.[6] When Gabriel was 11 years old, his mother filed taxes for his work as a child model—only to discover that his taxes had already been filed. It turned out that an illegal immigrant had adopted his Social Security number. From the opposite perspective, concealed identities can also be used for grooming unsuspecting child victims.

The extent of identity fraud is revealed in a 2017 study by Javelin Strategy and Research, which found that 6 percent of U.S. consumers had been victims of identity fraud in 2016, some 15 million in all.[7] The Insurance Information Institute gives a table of identity theft by U.S. state.[8] The top three (or perhaps that should be bottom three, in that they had the most identity theft) in 2016 were Michigan, with 176 complaints per 100,000 population; Florida, with 167; and Delaware, with 156. The state with the smallest number of complaints per 100,000 population was Hawaii, with 55.

Every year brings new reports of identity fraud. The Datashield website describes five of the "worst cases of identity theft ever."[9] Of course, whether a theft meets the criteria to be one of the worst five ever will depend on what those criteria are, but however you look at it, these cases are disturbing. For example, they include the case of Phillip Cummings, who stole 33,000 logins and passwords from his former employer, selling them to criminals who used them to steal between $50 and $100 million. They also include the case of father-of-two Malcolm Byrd, who was arrested for cocaine possession and spent time in jail before demonstrating to the police that his name had been stolen and used in the crime.

The essence of identity theft is impersonation, concealing your identity and adopting that of someone else and, usually, seeking to remain in the background as much as possible—perhaps never actually meeting the targets in person. But there are exceptions. The *Sunday Times* (London) of 4 March 2018 reported that the Zoosk dating website had images and a description of smiling gray-haired 58-year-old Danish American widower Martin. He was an attractive man and a good catch for women seeking a partner. Curiously, however, the man bore a striking resemblance to Christian, a divorcee described and pictured on Elite Singles, as well as to Sebastian, who appeared on Facebook. But it turned out that the resemblance was not surprising after all. The pictures and backstory were all of the same man, 46-year-old Steve Bastin. Steve was happily married and had never visited these websites. Someone had gone to considerable effort to collate his photographs and life story—easier, I suppose, than trying to create a consistent one from scratch. Several women fell for the ruse, at least for a while, and unfortunately such cons are not that rare. Reports regularly appear in the newspapers of people—usually women—who have fallen for them. They have often given the fraudsters very large sums of money—to pay for fictitious operations or to tide them over until an inheritance materializes (really!) or until an imaginary business transaction goes through.

You can think of identity frauds as the accessing and use of information about you that you wish to keep secret, that is, data that *should be* dark data, at least to everyone but you and the service you are using. It is when these dark data become visible—perhaps because they are stolen—that problems such as identity theft arise. So, dark data are not intrinsically bad. In chapter 9 we look at ways you can strengthen your security around the data you wish to keep secret, but signs that you might be the victim of identity theft include plastic card charges for goods or services you did not order, being sent credit cards you did not apply for, being notified that a credit score has been checked, as well as the more obvious ones like discovering money is missing from your account.

Nowadays, banks and other financial institutions have highly effective methods for detecting suspicious interactions. But they cannot be perfect—there is always the human element. A colleague of mine was confident that his new Mercedes would not be stolen. It had all the cutting-edge antitheft technology, ranging from secure electronic keys to cameras and a GPS tracker transmitter. But none of this helped him when he was mugged as he was getting into the vehicle.

I've deliberately focused on identity theft in these examples, but internet fraud comes in a huge variety of shapes and forms, all based on hiding information of some kind. One which you may well have encountered is "advance fee fraud."

Advance fee frauds are scams in which the victim receives an email promising them a large sum of money in exchange for facilitating some transaction. The facilitation will require the payment of some fee to cover initial costs like delivery or transaction charges. And then it typically turns out that further fees are needed, and so on—until the victim begins to suspect that it is a scam. Perhaps the most famous of these is the Nigerian 419 scam, named after a part of the Nigerian penal code. Typically, a victim is emailed with the request to help transfer a large sum of money from an African country. Estimates of the global amount lost to this scam vary, but they range up to $3 billion, all built on concealing and misrepresenting information.

Personal Financial Fraud

I used to do research in fraud detection, especially in the finance sector. At a conference of bankers, after I had given a talk on methods of detecting credit card fraud, one very senior banker came up to me and said, "We don't have any fraud at my bank." As I was thinking about this statement afterward the most generous explanation for his assertion I could come up with was that he was joking. But it was also possible that he was deliberately taking a corporate party line, not really expecting to be taken too literally. After all, if it became widely known that his bank suffered from fraud, perhaps suffered substantially, then it would damage the bank's reputation. There was good reason—at least for him, if less so for his customers—in keeping quiet about the fraud. That is, in hiding the data. Of course, there is also a third, perhaps more troubling, explanation for his assertion. This is that he was unaware of the fraud within his bank. Perhaps he genuinely believed that there was no fraud. That would mean that, at least as far as he was concerned, the data truly were hidden, and if that was the case, then it was worrying. The fact is that all banks suffer from fraud or attempted fraud.

The key role that dark data play in finance is nicely captured in Arjan Reurink's definition of financial fraud as "acts and statements through which financial market participants misinform or mislead other participants in the market by deliberately or recklessly providing them with false, incomplete, or manipulative information related to financial goods, services, or investment opportunities in a way that violates any kind of legal rule or law, be it a regulatory rule, statutory law, civil law, or criminal law."[10] Although Reurink is mainly talking about financial markets, the words "misinform," "mislead," "false," and "incomplete" are all characterizations of forms of dark data.

The range of different types of financial frauds is limited only by the imagination of those who wish to perpetrate them. But they are all based on misleading people about the underlying truth by concealing it. To see how some of these different types arise, let's begin with something personal: fraud with plastic cards such as credit cards and debit cards.

Plastic-card payment technologies continue to evolve. In the early days they were based on taking a print of the card, coupled with a physical signature. These were replaced by systems in which the user's details were encoded in a magnetic stripe on the card. Then, some ten years ago in Europe, and more recently in the United States, they changed to using a chip embedded on the card, in conjunction with a Personal Identification Number (PIN) memorized by the customer. More recently still, so-called contactless payment has been introduced for small transactions. This technology uses radio-frequency identification or near-field communication, which allows customers simply to tap their card or other device such as a smartphone onto a card reader. Since no proof is required that the person waving the card is the actual card owner, theft of the card means money (albeit in small amounts) can be stolen until the card is stopped—though only a limited number of such transactions can be made before a PIN is required.

Credit card numbers and PINs are example of kinds of data you will want to keep dark and to make visible only to the people or machines you choose. But those are the data thieves will be trying to read. The cards themselves do not need to be stolen, but only the information on the cards or the information used when making transactions with them. A mix of technology (e.g., "skimmers," card readers attached to real payment terminals which capture data from everyone who uses the terminal) and social engineering (e.g., tricking someone to divulge a PIN) is used by the thieves to access this information. Understanding these strategies equips you with the tools to decrease the chance you will fall prey to them. These tools include standard strategies like not letting anyone see your PIN as you enter it and not letting anyone take your card out of your sight.

There are various types of credit card fraud, and the importance of each changes over time as new prevention technologies come into force. The fundamental problem is that while new methods for detecting and preventing fraud may deter some people, they will not deter all, and certainly not organized crime. This means that the introduction of tools for deterring one kind of fraud can lead to an increase in other kinds of fraud. This "waterbed effect"—in which pushing crime down in one place leads to its rising elsewhere—certainly happened when chip-and-PIN technology was introduced to the UK before it was introduced elsewhere in Europe. The consequence of the UK introduction was that while crimes due to stolen cards went down in the UK, they went up in France as card details were sent across the Channel.

Among the most common kind of frauds are "card not present transactions." As the term suggests, these are transactions conducted remotely, over the internet or via phone or mail order, where the card and cardholder are not physically present. The riskier nature of such transactions explains why enhanced security procedures may be initiated if you wish to send a purchase made over the web to a new address. They are attempts by the vendor to overcome the invisibility of the person making the transaction.

Unfortunately, no matter how advanced the technology, human nature will prevent fraud from being stopped completely. You may have received an email purporting to come from a friend or colleague who is overseas and has had his money and papers stolen and needs cash to help him return home. Or you may have been a victim of a phishing attack, in which an apparently trustworthy email seeks to trick victims into revealing card information by luring them to log onto a fake website which looks exactly like their familiar bank or credit card site. Unfortunately, the battle between the fraudsters and those seeking to stop them is very much an arms war, with both sides becoming more advanced over time. Indeed, it was not so long ago that you could recognize many email phishing attempts by their poor spelling and inadequate grammar, but now they seem to have learned to spell. (Though it has been suggested to me that the poor spelling was deliberate, because anyone who was taken in by such semiliterate messages would be easier to steal from. Personally, I think that's giving the fraudsters too much credit.)

Each new wave of technology is intended to enhance user convenience, as well as improve security, but sometimes these are uneasy bedfellows. For example, the transition from magnetic stripe to embedded chip slightly slowed down transactions, sometimes frustrating customers in a hurry. Make a service too burdensome, and the potential customer will go elsewhere. This obstacle applies at several levels. Passwords, double-key identification, and biometrics such as fingerprint-, iris-, and voice recognition are all ways of keeping the account details hidden from potential fraudsters, of keeping your data dark. But they all inevitably add a burden to the use of the account. On top of the hurdles introduced in accessing his or her money, if a suspected fraud is detected, the account holder has to deal with the phone call asking if he or she made the suspicious transactions. This is all useful at some level—indeed, it can be reassuring knowing that your bank is keeping alert to protect you—but it becomes tedious if too many such calls are received.

Financial Market Fraud and Insider Trading

In 2011, Ghanaian financial trader Kweku Adoboli carried out off-the-book trades while working in the Global Synthetic Equities trading team at the UK office of the Swiss bank UBS. He ran up losses of some $2.3 billion. This was the largest unauthorized trading loss in British history, but not the largest ever. In the 1990s, Yasuo Hamanak, one-time chief copper trader at Sumitomo, carried out unauthorized trades which led to the company's losing $2.6 billion. And unfortunately there have been even larger losses still, though not all due to dark data and crime. Howard Hubler III undertook legitimate but risky trades in subprime mortgages in the United States in the early 2000s, causing Morgan Stanley to lose around $9 billion. It's in the nature of chance and risk that sometimes things go bad—though Hubler did hide information and misled his colleagues by telling them the financial positions they were taking were very secure.

While it is clear that some rogue traders set out to be fraudulent from the start, it seems likely that most do not. They might begin by carrying out trades which exceed authorized limits, driven by the organization's culture to make as much money as possible. Then, when they start to lose money, instead of admitting it and unwinding their positions at a loss, they may extend their positions in the hope that things will come out all right and no one will discover their unauthorized trades. The risks accumulate, and things can accelerate, increasing the pressure on them to make more overtly fraudulent trades. And from there it's all downhill. The activities of Nick Leeson, whose rogue trading activities leading to a $1 billion loss single-handedly brought down the 200-year-old British investment bank Barings, are a case in point.

All this talk of billions might be desensitizing you to the sheer magnitude of such numbers—as reflected in the humorous quote often misattributed to Illinois Senator Everett Dirksen to the effect that "a billion here and a billion there, and soon you're talking about real money." So let's put $1 billion into perspective. According to the Census Bureau, the median U.S. personal income in 2016 was $31,099. Thus, the $9 billion Morgan Stanley loss represented the annual income of nearly 300,000 people.

The phrase *insider trading* refers to using confidential information to take unfair advantage when trading in financial securities on a stock exchange. "Confidential" means the information is not known to the general public, so it is dark data. It is also asymmetric information (*DD-Type 12: Information Asymmetry*) in that it is known to one party to a transaction but not to the other, as we discussed in chapter 5.

As you might imagine, detecting insider trading can be difficult. The trick is to spot unusual patterns of behavior, such as people making multiple suspiciously well-timed trades in advance of public announcements.

One of the most famous insider trading scandals involved the American trader Ivan Boesky. In 1975 he launched Ivan F. Boesky & Company, specializing in speculating on corporate takeovers. He was extremely successful, making some $200 million within 10 years and being featured on the cover of *Time* magazine. But when Boesky successfully predicted most of the major deals of the 1980s, the Securities and Exchange Commission (SEC) became interested. Boesky's trades seemed remarkably well timed, with purchases of stock often occurring just before the stock value rocketed in response to a takeover. It turned out that he was able to do this not because he had precognitive abilities or because he had some superbly predictive algorithm but simply because he paid employees of investment banks to feed him information on impending takeovers. Boesky was taking advantage of data which should have been concealed. He was fined $100 million and went to prison. Apparently, the character of Gordon Gekko (he of the famous "greed is good" speech) in the film *Wall Street* is based partly on him.

Not all insider trader frauds involve sums as large as Boesky's. Australian Rene Rivkin based his decision to buy 50,000 Qantas shares on a confidential conversation with Gerry McGowan, chairman of Impulse Airlines, in which he learned that Qantas would merge with Impulse. But Rivkin ended up making a profit of just $2,665 (that's dollars, not millions of dollars). Of course, failure to succeed in making millions of dollars from a fraud is not a mitigating factor, and unfortunately for Rivkin, he was later found guilty of insider trading and sentenced to nine months in prison. He committed suicide in 2005. Later investigation revealed that despite being banned from trading, Rivkin had continued to do so in secret.

The crimes I have just described occurred before the so-called big data and data science revolution, so authorities often had to rely on being alerted to suspicious behavior by whistleblowers, other regulatory bodies, or financial exchanges. But as we advance into the big data world, modern machine-learning and artificial intelligence algorithms are proving invaluable in spotting unusual behavior and detecting concealed activity. In 2010 the U.S. SEC set up the Analysis and Detection Center, tasked with analyzing billions of trading records to spot unusual trading behavior.

The activities of the center have resulted in various charges being brought. For example, based on the center's work, in

September 2015 the SEC charged two lawyers and an accountant with purchasing shares in New Jersey–based pharmaceutical company Pharmasset after having been fed confidential information by a board member that the board was negotiating to sell the company. These three and two others agreed to pay nearly half a million dollars to settle the charges.[11]

Insider trading hinges on knowing something others don't. A more general manifestation of such asymmetric information occurs in false accounting. Hiding the true state of a company by concealing data and releasing incorrect accounting information is probably one of the most prevalent manifestations of dark data in the world of financial fraud. It might be false information about the real condition of an investment or potential investment, it might be concealment of improper transactions in order to mislead investors or regulators, it might be false information about earnings or profits, or it might be a host of other lies.

There are unfortunately many examples to choose from. A well-known case is that of Enron, one of the largest-ever corporate bankruptcies, which occurred in 2001. (Indeed, it was *the* largest corporate bankruptcy until the collapse of WorldCom the following year.) Enron was created in 1985 in the merger of Houston Natural Gas and InterNorth, and Kenneth Lay, previously CEO of the former, soon became the CEO. Enron became one of the world's biggest energy, communications, and pulp and paper companies, with revenues in excess of $100 billion. The complexity of the company enabled its chief operating officer, Jeffrey Skilling, and its chief financial officer, Andrew Fastow, to use accounting loopholes and specially set up companies to isolate Enron from financial risk, essentially hiding billions of dollars of debt from the board of directors. But in 2001 an article in *Fortune* magazine suggested it was not clear how Enron was earning its income and hence how it could justify a stock value some

55 times its earnings. Gradually, other concerns surfaced—not helped by Skilling's being gratuitously offensive to a journalist in a recorded conference call. Skilling resigned, initially citing personal reasons but then admitting that his move was motivated by a 50 percent drop in the company's share price.

On 15 August 2001, an anonymous letter from Sherron Watkins, the vice president for corporate development, warned Kenneth Lay about dubious accounting practices, saying, "I am incredibly nervous that we will implode in a wave of accounting scandals." That statement turned out to be prophetic. The company struggled on, but investor confidence was suffering, with various media attacks and concern about the opacity of its operations. Enron's share price fell from $90.75 in mid-2000 to $1 by November 2001 (prompting a $40 billion lawsuit from shareholders). Eventually, after a hard-fought battle by Lay, Enron's credit rating was reduced to junk status, and it filed for bankruptcy.

One might have hoped that events such as this would lead to increased regulation, making it more difficult to conceal the way companies operate. Unfortunately, this appears not to be the case. A 2014 article in the *Economist* alarmingly said: "If accounting scandals no longer dominate headlines as they did when Enron and WorldCom imploded in 2001–02, that is not because they have vanished but because they have become routine."[12] The article went on to describe a few others: Bankia in Spain misstating its finances when it went public in 2011; the Japanese company Olympus hiding billions of dollars of losses; the failure of Colonial Bank in 2008; the Indian technology company Satyam faking over $1 billion of cash; and so on. It's clear that such activities occur throughout the world, and it's unlikely that they apply only to giant companies and frauds involving billions of dollars. If even massive incidents no longer capture the headlines like they used to, what does this say about the number of lesser incidents?

Insurance Fraud

Most of us will have no direct contact with money laundering and corporate fraud, but an area with which we all have contact and where financial fraud occurs all too frequently is insurance. Like the Roman god Janus, insurance fraud faces in two directions. One is defrauding insurance companies. The other is defrauding insurance customers. Both depend on hiding information, the difference being who is hiding it from whom. Both can be deliberate and planned or can simply be opportunistic. Sometimes, planned fraud and opportunistic fraud are called "hard" and "soft" fraud, respectively.

One way in which customers might be defrauded is by the payment of premiums on nonexistent policies—which they might not find out about until they submit a claim, and that might never happen. At a higher level, customers might even be sold policies from nonexistent companies. Clearly, dark data scams like this require organization and are unlikely to be spur-of-the-moment opportunistic efforts. Needless to say, the web provides an ideal environment for disguising the truth of such operations.

"Churning" is another general strategy for fraudulently extracting money from customers, and one which requires elaborate organization. Here an unnecessary or excessive series of trades or transactions are undertaken, with commissions being charged on each one. In insurance, this might involve multiple intermediaries, each taking a commission. When looked at in isolation, each of the individual transactions looks unobjectionable, and it is only when the big picture is seen that the fraud can be detected. In fact, the difficulty of detecting fraud when each component of a network appears to be (indeed, is) behaving legitimately applies more generally.

A major consumer bank once employed me as a consultant to develop tools for detecting fraudulent networks in mortgage applications: if the various players are working hand in glove to manipulate prices it can be difficult to recognize that fraud is going on (though be warned, should you be tempted: modern data mining tools to ferret out deception are becoming increasingly sophisticated, and detection is very likely!).

Fraud in the other direction, when customers try to steal from insurance companies, is probably more common. Often, a policyholder misrepresents material facts, such as previous claims history, preexisting medical conditions, or vehicle modifications, so as to lower insurance premiums. A familiar example is insuring a property for more than it is worth before burning it down. Clearly, such scams require some planning, or at least forethought.

An even more extreme example is that of people faking their own death or the deaths of others so they can make fraudulent life insurance claims. Elizabeth Greenwood, who has written a book about such scams, estimates there are hundreds of cases of this sort each year.[13] Often, such incidents are contrived to happen when people are traveling to countries where it is easier to obtain fraudulent death certificates. For example, Jose Lantigua from Jacksonville, Florida, tried to escape his financial problems by collecting $6.6 million in life insurance by faking his death in Venezuela.[14] He was later arrested in North Carolina, having adopted a different name. A less ambitious attempt was made by a mother and son from the UK who tried to claim a £140,000 life insurance payout by claiming that the mother had died in a road accident while they were on holiday in Zanzibar.[15] The mother moved to Canada, but insurance investigators caught up with her after becoming suspicious that they could find no record of the death with the Foreign and Commonwealth Office.

Of course, faking your own death does come with the disadvantage that you do really have to disappear. As Greenwood says, the question is whether you can walk away from everyone and everything in your life, and also whether you can acquire a new identity.

A milder version of this sort of fraud occurs when people file a claim for ruined holidays arising from food poisoning. The sums may not be trivial—Paul Roberts and Deborah Briton from Liverpool in the UK falsely claimed £20,000 arising from two trips to Spain. Unfortunately, on social media Briton described the holidays as "two weeks of sun, laughter, fun and tears. Met up with all our lovely holiday friends who made our holiday fab" and "Back home after a fantastic holiday, my favourite so far." It is tempting to take such people aside and point out that the key to keeping data dark is just that—you have to keep them dark. Social media seem to be a very powerful tool for revealing such scams, which says something about the common sense of those attempting such frauds, or at least of those who are caught. (There may be many others, more sophisticated, who manage to keep things hidden.) In the present case both Roberts and Briton went to prison.

One carefully planned strategy which has attracted public attention in the UK in recent years has been seeking compensation payment for injuries claimed to be the result of a car accident, where the "accident" was deliberate—the so-called crash-for-cash scam. Sometimes, multiple passengers, or even phantom passengers, will each submit injury claims. Claiming for whiplash injury has been a particularly popular strategy, as this is easy to fake, and average payouts range from £1,500 to

£3,000. According to Tom Gardiner of Aviva Insurance, between 2005 and 2011 the number of road traffic accidents fell by 30 percent, but the number of claims for whiplash injury went up by 65 percent, which certainly arouses suspicion.[16]

Fake vehicle accidents are not restricted to the UK, and to draw out such fraudsters, in 1993 New Jersey fraud investigators rigged a dozen fake bus accidents, with all the "passengers" in fact being undercover agents.[17] It might seem a little farcical, but video recordings from one of these fake accidents showed 17 people rushing to board the bus after the "accident" but before the police arrived, so that they could later claim to have been injured in the accident. The case was aggravated by some doctors later billing for visits and treatments which had not been carried out. In all, more than 100 charges were brought. This episode gives a rather depressing impression of the fundamental nature of humanity!

Opportunistic claims made by people who had nothing to do with an incident are not restricted to traffic accidents. It is thought that some $6 billion worth of claims following Hurricane Katrina in 2005 were fraudulent.[18] And after the BP Deepwater Horizon oil spill in the Gulf of Mexico in 2010, more than 100 people were jailed for making fraudulent claims against the company. According to the *Financial Times*, "in 2013, BP estimated that it was paying out up to $100 million per week in fraudulent claims."[19]

There are many other variants of insurance fraud, to the extent that the total value of detected dishonest insurance claims in the UK in 2016 was £1.3 billion, involving some 125,000 claims, with a suspected similar amount undetected.[20] And according to the FBI, non–health insurance fraud costs over $40 billion per year. (The discrepancy between the UK and U.S. figures, even taking the different population sizes into account, is probably due to the use of rather different definitions, rather than implying a difference in average degree of honesty between the two countries!)

Whatever the kind of fraud, be it personal banking, insurance, or other, there is a general principle that the cost of preventing fraud should be balanced against the loss if it succeeds. There's little point in spending $1 billion to prevent a $1 loss. But there are elementary tools which will prevent the bulk of fraud, and they should be adopted. For example, in financial accounting, reconciliation is undertaken to ensure that funds leaving an account match the amount spent, to ascertain in particular that no money is going astray. This is a fundamental way to check that you are seeing all the data, and you probably do the same with your bank account, comparing your records of expenditure with a monthly bank statement (if you don't, you should). While discrepancies might sometimes arise because of time delays, unexplained discrepancies can be an indication of fraud. Likewise, systems of double-entry bookkeeping are used to ensure that things balance—that all transactions are visible and transparent. Such systems probably originated in fifteenth-century Italy: the 1494 book *Summa de arithmetica, geometria, proportioni et proportionalita* (Everything about arithmetic, geometry and proportion) by Luca Pacioli is thought to be the first published book describing the system, which shows its long history.

As far as insurance fraud is concerned, indications hinting at possible concealed data include the submission of many claims, patterns in the kinds of claims submitted, a claimant who is very calm after submitting a large claim, handwritten receipts for lost or stolen goods, increased insurance just before a claim, and medical claims by seasonal employees about to stop work. Clearly, these signs are specific to insurance fraud, and while awareness of them might not be helpful in detecting fraud in other contexts, this is where the *DD-Types* of dark data can be useful. As illustrated in chapter 10, these show not how specific instances can arise but, rather, describe the different higher-level characteristics of dark data.

And More

Money laundering is the process by which illegally obtained money is sanitized to disguise its origins. These origins can be any illegal activity—drug peddling, slavery, illegal gambling, extortion, tax fraud, people trafficking, and others. These are all nefarious activities about which you want to hide data. A 2016 report from PwC (PricewaterhouseCoopers) estimated that money laundering transactions for $1 trillion to $2 trillion per year were being carried out globally.[21] That amounts to 2 to 5 percent of global GDP.

Money laundering involves three steps:

- "placement": feeding the money into the financial system;
- "layering": undertaking often elaborate financial transactions to make it difficult to trace the actual source of the money;
- "integration": using the money in legitimate ways so that it is merged with money from legitimate sources and appears clean.

The first two of these steps, especially the first, make use of dark data. Since the sudden appearance of a large sum of money in an account, with no obvious explanation of where it came from, is naturally suspicious, anti-money-laundering regulations require large transactions to be reported. For this reason, money-launderers typically split large sums into multiple small sums—for example, less than $10,000 if this is the reporting limit. This practice of splitting the money to keep it invisible to the regulatory radar is known as *smurfing*.

Businesses which receive a significant part of their income legitimately as cash can be used to feed illegally obtained funds into the system, simply by adding the latter to the former and claiming it was all obtained legitimately. The sorts of businesses where this is easiest to accomplish are service industries such as restaurants, casinos, bars, and manual car washes. The shift away from cash toward contactless electronic transactions is likely to make this harder to do by making transactions visible and traceable.

Gambling is another common strategy for placement. Even if the odds are against you on any particular bet, over time some bets will win, and they can then be reported as legitimate gambling winnings. The inevitable small percentage of loss on repeated bets can be thought of as the cost of having the money laundered.

Dark data also lie at the heart of an investment fraud tactic you may have heard of—the *Ponzi scheme*. Although it is named after Charles Ponzi, who used the method during the 1920s, he was certainly not the first to come up with the idea. Indeed, the strategy was described by Charles Dickens in his novels *Martin Chuzzlewit* and *Little Dorrit*. Such schemes promise investors rich returns but in fact do not invest the money given to them at all; they simply feed back some small percentage of later investments to the earlier investors so that it looks as if profits are being made. Inevitably, at some point, the scheme comes tumbling down when it runs out of potential future investors or people start asking for their money back—possibly because of deteriorating economic conditions. Then, the cold light of reality reveals the true nature of the investment company and how it makes its investment decisions. The Madoff fraud, mentioned in chapter 1, was an example of a Ponzi scheme that fell apart when the financial crash of 2008 caused many investors to try to withdraw their investments, only to discover they had evaporated. Prevention of such schemes is facilitated by transparency and in particular by regulations requiring investors to be able to see what is being done with their money.

We have already looked at insider trading, but insider theft is a more general kind of fraud and one which is notoriously difficult to detect. Insider theft involves staff with access to accounts siphoning off money for their own use. These can be individuals who find themselves unable to resist the temptation when they realize they have control over large sums of money. In fact, such cases often occur when employees find themselves in financial difficulties and, unknown to their employer, "borrow" from them—fully intending to repay the money when things improve, only to find that things don't improve, and they get in deeper and deeper, typically ending in a prison cell.

But insider theft can also occur on a larger scale, as part of organized crime, and can stretch over years. I came across one elaborate, and sad, case of this kind. A poor student received a scholarship from a minor foundation, which funded his college fees and also paid the rent on his accommodations for him. On the student's graduation, the foundation helped him find a job in a bank. The student, hardworking and trustworthy, worked his way up in the bank until he reached a position where he controlled significant sums. The foundation then contacted him, requesting him to transfer a large amount of money to a named account—in what appeared to be an entirely legitimate transaction. And then the foundation, and the money, vanished—leaving the innocent and trusting employee to face the music.

The list of types of financial fraud hinging on dark data is endless, as is the particular form they can take. As well as those already discussed, others include tax evasion—illegally failing to declare tax which is owed (distinct from the tax avoidance which I mentioned in chapter 5, which uses elaborate strategies to legitimately avoid tax)—and boiler room scams, in which fraudsters cold-call potential investors offering them "wonderful bargains" in overpriced or worthless shares or bonds.

At some level all these different kinds of fraud involve hiding information. Because they are so diverse, a wide variety of different strategies are needed to tackle them. These range from the painstaking checking of detailed records, through sophisticated statistical approaches; to modeling typical customer behavior patterns, via machine-learning and data mining tools for detecting anomalous transactions; to the raising of flags indicating suspicion when certain kinds of transaction configurations occur. Regarding dark data, the moral is the familiar one: if something looks too good to be true, then it probably isn't true. It is probably hiding the truth.

Chapter 7

SCIENCE AND DARK DATA

The Nature of Discovery

The Nature of Science

Science is all about discovering the nature of things and how they work. It's all about shedding light on the unseen. But dark data also lie at the root of science in a very practical sense. Fundamental to the practice of science is the Popperian notion (after Karl Popper) of "testability" or "falsifiability." The basic idea is that you come up with a potential explanation—a theory, conjecture, or hypothesis—for the phenomenon being studied, and then you test that explanation by seeing how well its consequences or predictions match what actually occurs. Recast in our terms, we know what unseen data should look like if our theory is correct, and then experiments produce data which we can match against our predictions. If the theory's prediction does not match reality, as revealed by the data, then the theory is replaced, modified, or extended so that, as well as successfully predicting past observations, it also predicts the new ones. In a sense this is an example of *DD-Type 15: Extrapolating beyond Your Data*. But it is deliberate extrapolation, based on a theory, with the testing objective in mind.

At least until the scientific revolution (and doubtless since, though it is hoped less overtly) advances in understanding were retarded by a (typically subconscious) reluctance to collect data which might disprove a theory—the confirmation bias already mentioned. That is, advances were held back by an unwillingness to make dark data visible. After all, if you have a solid theory which has been accepted perhaps for centuries (like the miasma theory of disease, which held that epidemics were caused by a poisonous vapor given off by rotting matter and was believed across Europe, India, and China from ancient times to the nineteenth century), why would you go looking for data which contradicted it?

My favorite historical example of someone who spotted this problem is given by the seventeenth-century philosopher Francis Bacon, who wrote: "Human understanding when it has once adopted an opinion . . . draws all things else to support and agree with it. And though there be a greater number and weight of instances to be found on the other side, yet these it either neglects and despises, or else by some distinction sets aside and rejects." Bacon illustrated the dangers of neglecting data with the story of a man presented with a picture of people who had escaped drowning in a shipwreck after praying, and who was then invited to acknowledge the power of prayer. Bacon asked where were the pictures of the men who were drowned despite praying.

A classic and colorful example of collecting data to test a theory is the observational evidence collected by Sir Arthur Eddington and Frank Dyson to test Einstein's theory of general relativity. Einstein's theory predicted that the path of rays of light would be bent when they passed close to massive objects. An ideal such massive object would be our sun, when bending of light would be revealed by an apparent shifting of the positions of stars whose rays passed close to the sun on their way to Earth. The trouble is that the brightness of the sun would swamp the starlight—except, that is, if the sun's light was blocked out by the moon. With this in mind, in 1919 Sir Arthur Eddington led an expedition to Principe, an island off the west coast of Africa, and Frank Dyson led one to Brazil, to observe the May 29th solar eclipse. Their photographs and measurements of the stars' apparent positions during the eclipse confirmed Einstein's predictions. This meant that Newtonian mechanics was a simplification or approximation, and Einstein's theory of general relativity provided a better description of reality. This really was an example of shining a light!

There is one particularly important implication from this description of how science works, namely, *you can never be sure that you have found the "real" mechanism of a phenomenon.* Science provides explanations, each becoming more powerful as understanding progresses, but it is always possible that any of those explanations might be proven wrong by new experimental evidence—though the more past experimental evidence a new theory must also explain, the tougher this becomes. This "contingent" nature, allowing theory to change as more data become available, is what distinguishes science from, for example, religion, which is not driven by evidence. Having said that, for convenience I shall occasionally refer to a scientific theory as "true" or "correct," but bear in mind that theories are always subject to contradiction by further evidence which leads to their being changed.

Thus, science is a *process*. In particular, it is not simply a collection of known facts—even though it is often taught as if it were, and even though it is often described in that way for simplicity. When children begin to learn about science in school, for example, they are taught things like the periodic table of elements, Newton's laws, and how rainbows form, rather than the careful notion of testing observation against ideas. While it is necessary that children should understand the world around them, this is perhaps a pity: science education should serve as the natural cradle of critical thinking in our schools, equipping the adults of the future to better evaluate what they are told, as well as teaching them facts.

The use of the falsifiability criterion—the idea of deducing the consequences of a theory and then comparing those predicted consequences with actual data—goes back a long way. The ancient belief that heavier objects fall faster than lighter ones is easily debunked by observation—as Galileo is purported to have shown by dropping balls of different masses from the Leaning Tower of Pisa and noting that they hit the ground at the same time.

Similarly, the notion that Earth is flat is one which superficially seems to fit the facts. After all, if you go on a long road trip, while there may be hills and valleys, on the grand scale you have no sense that you are driving over a curved surface. But more data, more evidence, shows that things are not so straightforward, and people have been aware of this for millennia. One piece of evidence came from watching sailing ships float off into the distance and disappearing, beginning with the hull and ending with the tip of the mast.

In summary, the fundamental process of science is one of testing theories against observed but previously dark data, where a mismatch between the theory and those data leads us to reject or amend the theory. But we need to recognize that a mismatch might have another explanation. If the theory and data don't match, it might be because there's something wrong with the data. If there's one thing I hope the examples in this book have made clear, it is that data are always at risk of errors, measurement inaccuracies, sample distortions, and a host of other problems, so this is a real possibility. This is why scientists put so much effort into producing accurate measuring instruments and taking measurements under precisely controlled conditions, whether they are concerned with measuring mass, length, time, intergalactic distances, intelligence, opinion, well-being, GDP, unemployment, inflation, or anything else. Accurate, reliable, and trustworthy data are a prerequisite for sound science.

The testability criterion is what distinguishes science from pseudoscience. It's not difficult to come up with explanations (e.g., "it happened by magic"), but unless the explanation has been subjected to a rigorous set of tests, we should be suspicious. Moreover, a theory which appears to have an explanation for *any* possible outcome that could conceivably occur is useless—it is certainly not science. A theory of gravity which conjectured that things would sometimes fall down, and sometimes up, and sometimes sideways, and when they fell in any direction said it was simply to be expected, would be of no use whatsoever. In contrast, Newton's explanation that masses are attracted to each other, which is why a dropped object falls down, is properly scientific: its prediction can be tested. And then once we have observed from repeated observations that a theory is generally true, it becomes part of the canon of knowledge, and we can make predictions and build machines which rely on it.

An example of a theory which has been criticized on the grounds it is just too comprehensive to be scientific or useful is psychoanalysis. Indeed, as Frederick Crews eloquently demonstrated in his book *Freud: The Making of an Illusion*, its origin is a triumph of dark data, involving generalizations from highly selected groups (even Freud himself, a one-man sample), reluctance to collect disconfirming evidence, unwillingness to see what really was happening (at one point Crews comments "every stage magician hopes that his audience will consist of precisely such eyewitnesses as Freud"), and taking denials that things had occurred as evidence that they had in fact occurred (that "no means yes"). Perhaps the most revealing indicator is that Freud never admitted being wrong. A scientist not prepared to accept that his or her theory might be wrong hardly conforms to the principle of potential falsifiability and surely cannot be called a *scientist*. Freud's own admission that "I am actually not at all a man of science, not an observer, not an experimenter, not a thinker. I am by temperament nothing but a conquistador—an adventurer" really says it all.[1] Or perhaps Freud's disclaimer means that at least some of the problem lies not with him but with those who have uncritically regarded one man's opinion as fact.

If Only I'd Known That

Since the scientific process consists of testing conjectured explanations against real data, it is hardly surprising that potential explanations often turn out to be wrong. If this were not the case, the entire enterprise would be simplicity itself. And while great scientists are immortalized because of the theories they came up with which turned out to be good explanations for empirical phenomena, this does not mean they don't also produce their share of mistaken theories along the way. Often, this is because something they don't know about is discovered, or because new data have been collected.

One of Charles Darwin's most outspoken critics was Sir William Thomson, who later became Lord Kelvin (after whom the temperature scale is named). He was one of the most eminent scientists of his day, becoming Professor of Mathematics at Cambridge at the age of 22. He is buried in Westminster Abbey with Isaac Newton (and, more recently, Stephen Hawking). Previous physicists had estimated how long the sun would last, based on the assumption that it was burning fossil fuels like coal, but Kelvin recognized that this would mean it would burn for only a few thousand years. So he developed a suggestion from Hermann von Helmholtz that it was gradually contracting, and that the gravitational energy released by the contraction was being converted into heat and light. Even so, he calculated, the sun could not possibly have kept burning for long enough for life to have evolved on Earth, so he argued that Darwin's theory of evolution did not fit the data.

But Kelvin was wrong. Crucial data—which became available only later—were missing from his argument. These were data showing that the mechanism behind the sun's energy was not chemical burning or gravity but a new kind of thing altogether: nuclear fusion.

Nuclear fusion is the process by which atomic nuclei are forced together so powerfully that they merge into a single larger atomic nucleus. In the process some mass is lost, and this mass is converted into energy, which is radiated away. The conversion factor is such that a tiny amount of mass transforms into an enormous amount of radiated energy, as is demonstrated by the detonation of a hydrogen bomb. The fuels for such reactions are deuterium (which is a hydrogen atom whose nucleus has a neutron as well as a proton; an ordinary hydrogen nucleus has no neutron) and radioactive tritium (a hydrogen atom whose nucleus contains two neutrons and a proton), which is produced in nuclear reactors by bombardment of the isotope lithium-6 with neutrons. The magnitude of the power which can be generated is illustrated by the fact that, theoretically, half a bathtub of water plus the lithium from one laptop battery would yield electricity equivalent to burning 40 tons of coal. Such a source of energy would solve humanity's energy problems, and we could do away with the pollution-generating fossil-fuel power stations: nuclear fusion reactions are "clean" in the sense that no radioactive waste is produced. This sort of reaction is the source of the sun's energy.

However, if we want to produce such a reaction on Earth, we encounter the difficulty that squeezing atoms together with the necessary pressure for them to merge requires a huge force and very high temperature. At the moment, the most effective way of achieving those is to put a layer of deuterium around a nuclear fission bomb. But nuclear fission bombs are clearly not a convenient or practical way of generating a well-regulated energy supply! Thus, several major research projects around the world are working on ways of generating the forces and temperatures required to produce controlled nuclear fusion and safely containing the resulting high-energy plasma. Since it would simply melt through any physical material, this plasma is balanced in a carefully created magnetic field which keeps it away from the walls of the physical container. Although such projects have been underway for many years, none have yet succeeded in generating more energy than is put into them. (Indeed, for this reason nuclear fusion is sometimes jokingly described as *always* being 30 years in the future.)

If Kelvin was mistaken because he did not know about nuclear fusion, others have been misled about fusion by faulty data. In 1989, two physicists, Martin Fleischmann and B. Stanley Pons, announced they had managed to achieve nuclear fusion without raising the raw materials to incredibly high temperatures—so-called cold fusion—simply by passing an electric current through a solution of lithium in deuterium oxide. Since deuterium oxide is a form of water (called heavy water), the potentially limitless supply of energy that this implied would revolutionize society.

This announcement naturally stimulated a great deal of interest, and laboratories all around the world tried to replicate the experiment. Some appeared to succeed—one in Moscow and one in Texas, for example—but most failed.

At a lecture at the Atomic Energy Establishment at Harwell in the UK a few days after the press announcement, someone asked Fleischmann if his experimental setup had been compared with control conditions. In Pons and Fleischmann's case, that would be an experiment run with ordinary water, in which the hydrogen in the water molecules does not contain a neutron. Strangely, Fleischmann refused to answer the question. Now, lack of an answer is suspicious (is it in fact dark data?). Certainly, if there were no results from a parallel experiment using ordinary water, then there were missing data—data which could have been crucial in shedding light on exactly what mechanism was generating Pons and Fleischman's results. Later, other criticisms of Pons and Fleischmann's original experiment were made, and as time progressed some of the purported replications were withdrawn. The consensus nowadays is that cold fusion is not a realistic proposition—although there are still some who hold out hope. It would, after all, represent a new dawn for humanity.

A lack of data also misled the chemist Linus Pauling, winner of a Nobel Prize in Chemistry and a Nobel Peace Prize. Arguably one of the greatest scientists of all time, Pauling contributed to an immense range of chemical and biochemical topics and published over a thousand papers. He was one of many scientists trying to determine the structure of DNA in the middle of the twentieth century. Studying electron microscope images led him to conjecture that the structure was likely to be helical. This was not altogether original for Pauling—over years of painstaking and meticulous work he had already convincingly demonstrated that helical structures existed in other molecules. Despite having no X-ray images or hard data on sizes and bonding angles between atoms, he suggested that DNA was a three-stranded structure. Although calculations showed that the positions he proposed for the atoms did not perfectly fit what data he had, he felt it was just a question of sorting out the details. Being very much aware of other teams of researchers working on the problem, especially at the Cavendish Laboratory in Cambridge, England, he was determined to publish first. So, on 31 December 1952, he and his colleague Robert Corey submitted a paper titled "A Proposed Structure for the Nucleic Acids" to the journal *Proceedings of the National Academy of Sciences.*

Two other researchers, Francis Crick and James Watson at the Cavendish Laboratory, had also previously conjectured that the structure was a triple helix but had rejected it on the basis of data provided by chemist and X-ray crystallographer Rosalind Franklin. When Crick wrote to Pauling pointing out some of the problems with the triple-helix model, Pauling did exactly what the scientific process described earlier implies he should do: he tried to modify his theory to match the data. But meanwhile, Crick and Watson were also looking for alternative models. And extra data from Jerry Donohue, a specialist in hydrogen bonding, led them to one such alternative which fit all the data—the famous double helix.

Pauling was not yet willing to accept that he had been wrong, writing that he looked forward to seeing which model would turn out to be right. When he traveled to Cambridge in April 1953, he looked over the structure Crick and Watson had come up with, and examined X-ray photographs. After discussing the model with them, Pauling acknowledged they appeared to have the answer.

It is in the very nature of science that even the most able and eminent scientists can get it wrong, especially when they don't have all the data. For example, Lord Kelvin, mentioned earlier, brilliant scientist though he was, made his fair share of other mistaken suggestions. When Wilhelm Röntgen announced the discovery of X-rays, Lord Kelvin's first reaction was that it must be a hoax. He also wrote that "no balloon and no aeroplane will ever be practically successful." And Albert Michelson (of the famous Michelson-Morley experiment which provided strong evidence in favor of Einstein's theory of special relativity) wrote in 1894, just before the discovery of quantum mechanics and relativity, that "it seems probable that most of the grand underlying principles [of physics] have been firmly established."

Sir Fred Hoyle is another eminent scientist who produced a theory which was refuted when his predictions were compared with data. Hoyle made major advances in our understanding of the universe, in particular, concerning the origin of heavy elements. A theory that the elements had formed early in the life of the universe had foundered when calculations showed that some of the steps in the process were too unstable to permit lighter elements to coalesce into heavier elements. Hoyle came up with an alternative explanation, showing how heavier elements could be synthesized in stars by fusion—exactly the sort of process we discussed earlier. Once synthesized in the cores of ancient stars, Hoyle argued, the elements were scattered throughout the universe when those stars exploded in giant supernovae. And then the elements slowly accreted to form planets, moons—and us. This theory has stood the test of time, establishing Hoyle as one of the most accomplished British physicists of the mid-twentieth century. But not all of Hoyle's ideas were so successful. The following is one such.

After data on the distance of stars from the Earth revealed that the universe is expanding, Belgian physicist Georges Lemaître made the logical deduction and suggested that perhaps the universe had begun as a tiny ultradense and hot point billions of years ago. Since the notion of testability is key to science, and there seemed to be no way to test this theory, and because it had no competitors, it did not attract much attention. But then Hoyle suggested an alternative to what he called Lemaître's "big bang" theory: that instead of having a well-defined beginning, the universe was in a state of continuous creation, with new matter continually popping into existence throughout the universe (the so-called steady-state theory). The existence of two competing theories stimulates the search for data to distinguish between them, since at least one of them must be wrong. In this case evidence gradually accumulated favoring the big bang theory. But Hoyle did not give up, devising many variants of his steady-state hypothesis to keep the theory in play. Eventually, however, the evidence against this theory became overwhelming.

Even Albert Einstein came up with theories which further data showed to be incorrect. His theory of general relativity showed that mass distorts space and time (which is what accounts for the bending of the paths of light rays as they pass a massive object, which we discussed earlier). At the time that Einstein described his general theory, the universe was thought to be static. However, since all matter attracts all other matter, a universe which just sits there would not sit there for very long: it would collapse on itself. To overcome this objection, Einstein added an extra term to his equations, a so-called cosmological constant, leading to a repulsive force which would counteract the attraction of gravity. Unfortunately for Einstein, this extra term proved unnecessary when further data revealed that the universe is not static after all but is expanding. Einstein is said to have called the introduction of this extra term his "biggest blunder." But he was being a little unfair on himself. After all, based on the data which were available at the time, conjecturing that such an extra force existed was a perfectly good idea. In general, if new data, previously dark and unsuspected, do not match a theory, it does not mean that the theory was a bad idea at the time. As it happens, however, there's another twist to Einstein's tale.

Further data collection has led to the discovery that not only is the universe expanding but that it is in fact expanding at an increasing rate. This has led to the suggestion that the cosmological constant, or at least something like it (nowadays it's called *dark energy*), might be necessary. Perhaps Einstein was right the first time. Incidentally, in his excellent book *Brilliant Blunders*, astrophysicist Mario Livio questions whether Einstein actually did use the phrase "biggest blunder." Livio attributes the phrase to physicist George Gamow.

Turning from pure science to medicine, we find that attempts to alleviate suffering have been with us since the dawn of humanity, with plants, earths, and magic all playing a role. Yet, it is only relatively recently that we have begun to properly evaluate the effectiveness of treatments, guided by a developing deep understanding of biology, physiology, genetics, and the associated sciences underpinning medicine. Thus, it is perhaps not surprising that some unevaluated treatments have lingered on in general practice. I'm not referring to things like the idea that someone might have too much blood and need bloodletting, or to discredited notions such as homeopathy, but to certain practices which the general medical community regards as effective but which have not been subjected to a rigorous evaluation, at least not until recently, such as through a randomized controlled trial.

The case of the prefrontal lobotomy is an apt illustration. This neurosurgical procedure was used around the world to treat psychiatric illnesses including schizophrenia and bipolar disorder for several decades. It involves cutting connections in the brain's prefrontal lobe. Initially this was done by drilling a hole in the skull and injecting ethanol to destroy a part of the brain, and later it was accomplished by inserting a rotating loop of wire. In still later developments, surgeons accessed the frontal lobes through the eye sockets. The inventor of the procedure, António Egas Moniz, was awarded the Nobel Prize for Medicine or Physiology in 1949 for devising it. This award was made despite suspicions that the procedure was ineffective: a 1941 editorial in the *Journal of the American Medical Association* said: "This operation should not be considered capable of transforming a psychotic personality into a normal one. Even in our present state of ignorance concerning the frontal lobes there is ample evidence of the serious defects produced by their removal in nonpsychotic persons."[2] Certainly, the procedure had significant side effects, including vomiting, bladder and bowel incontinence, lethargy, apathy, and a host of others; but on the other hand, treated patients were quieter and easier for families to look after. (The cyberneticist Norbert Wiener said: "Let me remark in passing that killing [such patients] makes their custodial care still easier."[3]) Fortunately, the use of the procedure began to decline in the middle of the last century as chemical medications were developed. Nowadays we have a much better understanding of the brain, and neurosurgical interventions are precise and carefully controlled exercises, enhanced by sophisticated scanning technologies. These technologies allow us to see within the three-dimensional structure of the brain—essentially making data visible.

A more recent medical example is arthroscopic surgery to ease osteoarthritis of the knee. This is a widely used procedure generally regarded as alleviating painful symptoms. But when Bruce

Moseley and his colleagues looked at its effectiveness in a randomized controlled trial comparing it with a placebo, they found that "at no point did either of the intervention groups report less pain or better function than the placebo group."[4] Blind assumptions about the effectiveness of treatments are unjustified—it is also necessary to make clear the consequences of not doing the treatment, so that a comparison can be made.

More generally, it is perhaps not surprising that we find many ineffective "treatments" on the fringes of medicine. For example, in a very recent major meta-analysis, Joonseok Kim and his colleagues found conclusively that "[multivitamin and mineral] supplementation does not improve cardiovascular outcomes in the general population."[5] But people are difficult to convince. Rather than accepting that their beliefs may be mistaken, they cast doubt on the evidence—confirmation bias again. Perhaps the most powerful example of this phenomenon at present regards climate change, but in the context of medicine, John Byrne says: "When faced with the hard reality of properly obtained, negative studies, it is natural for many to become denialists. Many will still cling to the favoured practices. Vitamin C is still given for colds. Cough medicines are actively recommended. Arthroscopic knee surgery is still performed (and avidly defended by some). Fenofibrate sales will likely still be in the billions for years to come. Being skeptical doctors means that we follow the evidence, even if we may not (at first) like where it leads. True compassion is guided by truth."[6] And truth is revealed by data.

Tripping over Dark Data

Sometimes we get lucky. Dark data usually represent a problem. Something is hidden from us—something which, if we knew about it, would change our understanding and most likely the actions we take. But occasionally we accidentally trip over dark data, and the world stands suddenly revealed before us.

A classic example was the discovery of cosmic microwave background radiation. The familiar story is that in 1948 Ralph Alpher and Robert Herman predicted the universe would be bathed in low-temperature radiation, a remnant of a period in the evolution of the universe soon after the big bang. Sixteen years later, in 1964, astronomers Arno Allan Penzias and Robert Woodrow Wilson were taking measurements with a 20-foot-long horn-shaped Dicke microwave radiometer. This instrument was originally built for testing satellite transmissions, but Penzias and Wilson were using it as a radio telescope. However, they had trouble eliminating background noise—even cooling the antenna did not remove it.

One explanation they investigated was the possibility that it might be caused by pigeon droppings from birds roosting inside it. They scrubbed it clean, but the interference remained. Coincidentally, Robert Dicke, who was looking for evidence to support the notion there would be fossil radiation left over from the big bang, visited their laboratory. On meeting Penzias and Wilson, he realized they had found what he was looking for, completely by accident while they were trying to do something else. Penzias and Wilson shared the Nobel Prize for Physics in 1978 with Pyotr Leonidovich Kapitsa for his work in the area of low-temperature physics.

The moral of this story is that while most anomalies and glitches in data will be experimental errors or measurement inaccuracies of some kind, others might lead to fundamental breakthroughs. The following are some more examples.

When B. C. Subba Rao tested 57 substances and found that one of them behaved in an anomalous way, he wanted to ignore that one result and publish the consistent results of the other 56.

This made sense, not least on the grounds that the 56 had been prepared in a carefully controlled way, while the 57th had been prepared in a different way. But his collaborator, English-born American chemist Herbert Brown, felt they should get to the bottom of things, and he investigated further. This led to Brown's discovery of a chemical reaction called hydroboration—and to his Nobel Prize.

German mechanical engineer and physicist William Röntgen noticed that when a high-voltage current was passed between two electrodes in an evacuated glass bulb, it caused a platinobarium screen 9 feet away to glow, even though the bulb was shielded with heavy black cardboard. He had discovered X-rays.

The planet Uranus was discovered by accident by German-born astronomer William Herschel when he noticed an object, barely visible to the eye, that changed position against the background of fixed stars.

Alexander Fleming is said to have discovered penicillin when, sorting through petri dishes containing colonies of the bacterium *Staphylococcus,* he noticed an area on one dish where there was a blob of mold but no bacteria.

In his seminal work *The Structure of Scientific Revolutions,* philosopher Thomas Kuhn wrote: "That is what fundamental novelties of fact and theory do. Produced inadvertently by a game played under one set of rules, their assimilation requires the elaboration of another set. After they have become parts of science, the enterprise . . . is never quite the same again."[7] But beware: while novelties, anomalies, surprises, or shocks may represent a flicker of light revealing previously concealed dark data, so that they give hints of a deeper understanding, they may also simply be a consequence of measurement inaccuracies or experimental glitches which are smearing data with an obscuring film.

Dark Data and the Big Picture

In a paper which has become famous, John Ioannidis, a professor of medicine and statistics at Stanford, claimed "it can be proven that most claimed research findings are false."[8] As a result of this bold assertion, he has become one of the most widely cited of scientists.

The reasons behind his claim are well understood, and indeed have been understood for decades. What was less appreciated until Ioannidis stirred up interest was just how big the effect was on the scientific and medical research literature. Also unexpected was the huge outpouring of interest and concern it precipitated. Perhaps even more surprising than all those outcomes, however, was that the subsequent debate revealed just how widespread were basic misconceptions about the very nature of the scientific process, often among those who should know better, with such questions being asked as, Why does science fail to meet the basic standard of reproducibility? Is science broken? and Is science really facing a reproducibility crisis?

The following are some figures to put Ioannidis's claim into context, bearing in mind that there are substantial differences between disciplines.

Bruce Booth, partner in the life sciences venture capital firm Atlas Venture, said that "an 'unspoken rule' among early stage venture capitalists is that at least 50 percent of published studies, even those in top-tier academic journals, 'can't be repeated with the same conclusions by an industrial lab.'" As a consequence, Atlas Venture now requires independent validation as a precondition for further investment.[9]

A survey by *Nature* revealed that over 70 percent of its 1,576 respondents had tried and failed to reproduce another's work.[10] We might wonder if this report should be taken at face value—it is likely that it has been distorted by dark data. For example, perhaps those scientists who had failed to replicate some work in the past would be more inclined to respond to the questionnaire than those who had succeeded. Or perhaps there was one study that produced anomalous results which many of the respondents had tried and failed to replicate (recall the case of cold fusion). Nevertheless, the figure is strikingly large.

In an extreme case, C. Glenn Begley and Lee M. Ellis describe studies aimed at reproducing the results in 53 "landmark" papers in preclinical cancer studies. They found that the findings were reproduced in only 6 of the 53 studies, that is, 11 percent.[11]

Leonard Freedman and his colleagues cited studies giving estimates of the amount of irreproducible preclinical medical research ranging from 51 percent to 89 percent. They then went further and estimated the financial cost of results that later studies could not replicate, finding that it came to some $28 billion per year.[12]

Motivated by results of this kind, in a famous investigation Brian Nosek from the University of Virginia led the Reproducibility Project, which sought to replicate 100 studies published in the psychological literature in 2008.[13] Of the 97 studies which had originally produced statistically significant* results, in only 35 of them did Nosek's team get the same result. As you might expect, this was not an uncontroversial study, with some researchers claiming that Nosek's design had major weaknesses, such as in its selection of the 100 studies to be replicated. It goes to show that even investigations of dark data can themselves be plagued by dark data issues. It is no exaggeration to say that dark data are everywhere.

*This is a technical statistical term which we define later.

These sorts of conclusions certainly seem alarming, but we must never forget that science is a winnowing process. Many of the critics seem to have an idealized child's view of science, in which an experiment is a one-off exercise which seeks to "prove" or "disprove" the existence of some phenomenon. Science is more complicated than that. And it has to be so. By definition, scientific research takes place at the boundaries of knowledge, where uncertainty dominates. Because researchers are trying to separate a tiny signal from a mass of noise, it is only to be expected that often the noise will tip things in the wrong direction. Indeed, we might go further and say that if we did not see experimental results which failed the reproducibility test, then scientists would not be doing their job. They would not be adventurous or creative enough in testing the boundaries of understanding.

The point of this discussion is that the scientific process is *not* broken. The irreproducibility is a sign that it is working, that claims are being tested and that those which are untrue are ultimately being rejected. Furthermore, the bottom line is that science *does* demonstrably work. We have only to look at our increasing understanding of nature and the advanced technology surrounding us, in materials, machines, and medicine to see this.

Admittedly, even if science is not broken, it would obviously be better if fewer incorrect conclusions were initially drawn—provided correct scientific theories weren't too frequently rejected. And the route to that lies in better design of studies. But there are also aspects of scientific culture which tend to encourage risk-taking and lead people to unnecessarily overstep the boundary, and this is what Ioannidis and the many who came after have stressed. We'll return to consider these matters in a moment.

First, however, let us note an analogy with the U.S. space program. In its early days, when rocket engines and systems were being refined and when the boundaries of our understanding of how rockets behaved were being pushed back, failures naturally occurred quite often. Or as Tom Wolfe described it in chapter 10 of his book *The Right Stuff*, the pessimistic view was "our rockets always blow up." Whenever you are exploring a boundary, you sometimes step this side of it and sometimes that—perhaps half the time each side if you are sitting right on that boundary. And from these results ("failures" perhaps in rocket engineering, but just "results" in science), through careful engineering you try to ensure that you land on the right side in the future: that your rockets don't blow up and that your scientific conclusions are correct. But the very nature of the enterprise means that's the risk you are taking. It's the risk you *have to take* to learn where the boundary is.

What aspects of scientific culture might drive researchers to results on the wrong side of the boundary, to spurious claims and irreproducible results?

Publication bias describes the fact that the published literature is not a fair representation of all the scientific work that is carried out. It means that large quantities of scientific results lie moldering away unpublished and undiscovered "in the drawers of filing cabinets," or gather metaphorical dust on hard drives that no one looks at. These hidden results are not simply a random selection but are selected by various forces (*DD-Type 3: Choosing Just Some Cases*). One force is the tendency of journals to prefer striking and novel results. An experiment which produced an unexpected outcome is likely to stimulate much more interest than one about which people simply say "ah, yes, that's what I would have expected."

This "file-drawer" effect stimulated the creation of the *Journal of Non-Significant Differences*. "Significance" is a technical statistical term which we shall examine in more detail later, but roughly speaking a significant result is one that is very unlikely to occur if a certain hypothesis is true, so perhaps leading to doubt about the truth of that hypothesis. And contrariwise, a nonsignificant result is one that is in accordance with the hypothesis: in the words of the preceding paragraph, it's what you would expect if the hypothesis is true. The *Journal of Non-Significant Differences* publishes only research which yields nonsignificant results, with the aim of communicating "an understanding that research does not have to be significant to provide valuable insight into ongoing scholarship."[14] (Incidentally, there is also a *Journal of Irreproducible Results*, founded in 1955, but that is a science humor magazine.[15] It's important not to confuse the two!)

Why might the preference for the novel and unusual lead to irreproducible results? Well, extreme, abnormal, or anomalous values can arise because of a genuine underlying phenomenon: perhaps this drug really is much more effective than the others, or perhaps this alloy really is far less reactive than expected, for example. But such values can also arise purely from random fluctuation. A chance coincidence of background conditions (impurities in the drug or alloy; participants in a psychology experiment feeling under the weather) or simple measurement error might lead to a spuriously high or low value. After all, as we have seen, no measurements are *absolutely* accurate, and no two experimental setups or samples are completely identical.

It follows that it's unlikely that the same accidental configuration or error of measurement will appear when the work is replicated. On average, as we saw in chapter 3 when we examined regression to the mean, we should expect later replications to produce values which are less extreme. We should not be surprised if the anomalous result disappears, and we should expect that often the "claimed research findings are false," as Ioannidis put it. In our terms, the unusually extreme values arise because of a form of dark data—the spurious measurement error—which serves to distort the underlying value.

The tendency of the editors of scientific journals to prefer exciting results is just the end point. Earlier in the chain, authors will be less inclined to submit papers they believe have a poor chance of being accepted for publication, with an interesting and rather surprising consequence.

There is tremendous prestige attached to publishing in high-profile journals which have a high impact factor—such as *Nature* or *Science*, to name just two. (*Impact factor* is a measure often regarded as indicating the importance of a journal. It's based on numbers of citations of the articles in the journal, that is, on how much attention has been paid to them.) Researchers are more likely to submit striking "breakthrough" results to high-prestige journals than they are to submit more mundane run-of-the-mill results, simply because they know the former journals prefer to publish such results (*DD-Type 4: Self-Selection*). This in turn means these journals tend to publish a higher rate of the more exciting advances—which is one of the reasons for their high prestige in the first place (*DD-Type 11: Feedback and Gaming*). But more outstanding results are also more likely to be unreplicable, owing to the regression to the mean phenomenon we have just discussed. Additionally, researchers might be more prepared to "bend" the precise details of what they did to increase their chance of being published in such places. (As we shall see later, there is often a fine line between legitimate and fraudulent selection of data points). The implication would be that the more prestigious journals are likely to have more fraudulent results.

The surprising consequence seems to be that the results published in leading journals are less likely to be true. And this is indeed what various authors appear to have found. Ferric Fang and coworkers have reported that "a highly significant correlation was found between the journal impact factor and the number of retractions for fraud or suspected fraud and error."[16]

A tongue-in-cheek conclusion is that to avoid publishing in outlets which have a high rate of false results, you should avoid high-prestige journals. It's all very confusing! But we should also bear in mind that causal directions are difficult to establish. The higher-profile journals, by definition, have a wider readership, so their papers naturally come under closer scrutiny, with the implication that questionable conclusions are more likely to be detected, even if the underlying rate is no higher than in other journals.

The problem is alleviated if announcements of novel results are required to be based on more than one study, that is, if independent replications are conducted before announcements are made (recall Bruce Booth's "unspoken rule"). This is the case in the pharmaceutical sector, for example, where drugs submitted for regulatory approval have to be backed up by multiple clinical trials. But in other contexts—especially in academic research environments—priority of publication is a key measure of performance, and researchers are loath to hang around risking that others will get there before them (recall Linus Pauling's rush to get his paper on the structure of nucleic acids published). It's better to publish a result which might be a major breakthrough, even if there is a risk that it will be found to be flawed.

The pressure to make novel discoveries has led to the practice of looking at data in many different ways, or looking at many different parts of a data set, until something significant is found. For example, in comparing two groups of patients, we might measure 100 characteristics of each patient and then compare the averages of the two groups on all 100 characteristics, one at a time. It would be surprising if the groups were not substantially different on at least some of the characteristics, purely because of random measurement error. This procedure is sometimes called *p-hacking*.

The curious name comes from statistics.

To begin with we need to recognize that if we look at data, especially large data sets, in enough different ways we are essentially certain to find some unusual patterns just by chance, even if there are no real relationships underlying the data. For example, if a great many variables are involved, then some pairs are likely to have high correlations in any finite set of data even if the real underlying variables are not correlated, just because of randomness and measurement error. Or if a great many cases (e.g., people) are involved, then there are very likely to be some small groups which happen, by chance, to be surprisingly similar.

To take a trivial but straightforward example, suppose we produce a string of 1,000 randomly chosen digits. Here are the first 30 of such a string:

678941996454663584958839614115

Now, we might look for a substring of 10 consecutive occurrences of the same digit among these 1,000. If we are unsuccessful, we might instead look for a sequence of nine consecutive digits occurring in order (123456789). If that fails, we might look for a sequence like 2323232323, of alternating digits. If that fails . . . Eventually, if we keep trying for long enough, we can be sure of finding some structure in the data. But here's the crunch: the structure, the anomaly we find, will not represent anything real. If we were to repeat the exercise, generating another 1,000 digits at random, there's no reason to suppose that we would find the same unusual configuration among them. The discovery would not be replicable.

Economist Ronald Coase has pithily summed up this situation, saying that if you torture data for long enough they will confess. But like other admissions extracted by torture, the confession may not represent the truth. In our example, the 1,000 digits were generated at random—there is no hidden meaning behind any structures we might find.

The concept of p-hacking formalizes this situation. We begin by noting that a fundamental tool in scientific research is the *significance test*. This is a formal statistical procedure to test a hypothesis or conjecture. We start by calculating a summary statistic for our sample. For example, we might choose to summarize the sample using the mean or median or variance, depending on what aspect of the data we are interested in. Now, if we take another sample, it is likely that it will have a different value for the summary statistic. Indeed, if we were to repeatedly draw samples, we would obtain a distribution of sample statistic values. Statistical methods enable us to calculate the shape of this distribution under the assumption that our hypothesis is true.

Then, by comparing the actual observed value of the summary statistic with this distribution, we can see how often we would *expect* to get a sample summary statistic at least as extreme as that obtained from the real data if the hypothesis was true. The probability that we would get a summary statistic value at least as extreme as that actually observed is called the p-value of the test. A very low value—say p equal to 1 percent—would mean that if our hypothesis was right, then we'd get a sample as extreme or more extreme than our actual data only 1 in 100 times. That suggests that either our hypothesis was right and a very improbable event has occurred, or the hypothesis was wrong.

It is often convenient to compare p-values with conventional thresholds. If our analysis produces a p-value no greater than the threshold, then the result is said to be *statistically significant* at that threshold level. For example, if we chose a threshold of 5 percent, then p-values this size or smaller would allow us to say that the result was "significant at the 5 percent level."

Here's an example. I might hypothesize that a particular coin is fair; that is, the probability of getting a head each time I toss it is a half. To test this, I'll toss it many times and see what proportion comes up heads. Under the hypothesis that it is a fair coin, I'd expect about half of the tosses to come up heads. I wouldn't really expect to get *exactly* equal numbers of heads and tails; I'd expect *some* slight difference from 50 percent heads, but I wouldn't expect an extreme difference if it was a fair coin. The significance test tells us the probability, *if the coin is fair*, that we'd see a difference the same as or greater than the difference we actually obtained when we did the experiment. Obviously, if it is very unlikely that such extreme results would be obtained were the hypothesis true, we might be suspicious about the hypothesis. For example, the chance of getting 90 or more heads in 100 tosses with a fair coin is astronomically small (this chance would be the p-value). So, if the coin came up with 90 heads when we tossed it 100 times, we might suspect that the coin was not fair.

Incidentally, the p-value is a horribly misunderstood concept. People often think it gives the probability that the hypothesis is right. It doesn't. The hypothesis is either right or wrong, and the p-value simply tells you the probability of getting certain extreme outcomes if it is right.

So that's what a p-value is, but where does the "hacking" come from?

The term arises from the misleading practice of carrying out lots and lots of significance tests without taking the number of tests into account, and it's easy to see why this is a problem. Suppose we tested 100 unrelated hypotheses, all of which, unknown to us, were true. And suppose we treated a p-value of 2 percent for any one of these 100 hypotheses as being sufficiently low that we would be suspicious about that hypothesis. For any one of the significance tests alone, that's perfectly reasonable. It means the chance of incorrectly raising suspicions about that single hypothesis if it is true is only 2 percent. But with 100 such tests, each carried out at the 2 percent level, it turns out that the chance of raising suspicions about *at least one of them* is 87 percent. You are more likely than not to decide that at least one of the hypotheses is suspicious, *even though all are correct.* Recall the comment about torturing the data for long enough! If you hide the fact you have carried out 100 tests, treating this as dark data (*DD-Type 2: Data We Don't Know Are Missing*), then the conclusions may be very misleading.

This sort of blunder has been made time after time in the scientific literature. A 1987 survey of randomized trials from four leading medical journals found that "74 percent of all trials were judged as having at least one impaired comparison, and 60 percent had at least one emphasized comparison that was impaired by the statistical problem of multiple comparisons. None of the trials judged as having impaired comparisons discussed the potential impact of the multiple comparisons problem on their conclusions." By "impaired by . . . multiple comparisons" the authors meant that the studies failed to take into account the multiple statistical tests being run, so that false positives were highly likely. We might hope the problem has been reduced since that study, but experience suggests it is still insufficiently recognized.[17]

A classic paper on this problem is much more entertaining than its title might suggest: "Neural correlates of interspecies perspective taking in the post-mortem Atlantic Salmon: an argument for proper multiple comparisons correction," by Craig Bennett and his colleagues.[18] This study looked at the results of an MRI scan of a dead salmon's brain when "the salmon was shown a series of photographs depicting human individuals in social situations. . . . The salmon was asked to determine which emotion the individual in the photo must have been experiencing." You can probably draw your own conclusion about the response of the dead salmon's brain to being shown these photographs. But MRI scans are made up of about 130,000 elements (called *voxels*), each one of which has a tiny chance of showing electrical activity purely because of random background noise in the equipment, and not because the cells in the dead salmon's brain were really responding. Now, while it's true that *each* of these voxels has only a tiny chance of showing a spurious signal, there are a *great many* of them. Add up that vast number of tiny chances and it becomes overwhelmingly likely that *at least one and probably more* of those voxels will appear to show electrical activity, giving the appearance that some neurons in the salmon's brain are firing, even though the salmon is dead. And Bennett et al. did indeed find apparent signals from some of the voxels. They concluded: "Either we have stumbled onto a rather amazing discovery in terms of post-mortem ichthyological cognition, or there is something a bit off with regard to our uncorrected statistical approach. Could we conclude from these data that the salmon is engaging in the perspective-taking task? Certainly not. By controlling for the cognitive ability of the subject we have thoroughly eliminated that possibility."

Bennett's paper won an Ig Nobel Prize in 2012. This award is given to "honor achievements that first make people laugh, and then make them think."

I am reminded of the following joke. Experimenter A tells B that he's having great trouble in replicating B's results. "I'm not surprised," says B, "I didn't get it the first hundred times I ran the experiment either."

Sergio Della Sala and Roberto Cubelli[19] have drawn attention to what might be another example of p-hacking, in which Randel Swanson and colleagues report on possible brain damage suffered by U.S. diplomats in Havana after alleged exposure to "an unknown energy source associated with auditory and sensory phenomena."[20] Swanson et al. concluded that "these individuals appeared to have sustained injury to widespread brain networks without an associated history of head trauma."

But how did they go about testing this? Table 2 of the Supplement to Swanson's paper describes their having looked at 37 neuropsychological test results, and a footnote to the table says "*Bold highlighting denotes abnormality or <40th percentile.*" That seems to imply that anyone scoring below the 40th percentile of the distribution of scores on any of the tests will be categorized as "abnormal." That is certainly how Della Sala and Cubelli interpreted it. But if the 37 tests were perfectly correlated (all giving the same result for any chosen individual), this would mean that only 60 percent of the population would score above the 40th percentile threshold on all tests and hence be categorized as normal. And at the other extreme, if the test scores were perfectly independent, with no correlations between them, then a straightforward calculation shows that fewer than one person in a hundred million would be categorized as normal on all the tests. Superficially, it appears that Swanson et al. bent over backward to show there was some brain damage in at least some of the people. As Della Sala and Cubelli point out, it would have been far better to choose a much more stringent criterion for abnormality—say 5 percent, instead of the 40 percent which they appeared to use—but a key problem is allowing a score below the threshold on any *one* of the 37 tests to suggest abnormality.

Incidentally, so that there's no misunderstanding: none of this means that these individuals did *not* suffer brain injury. It merely means that the conclusion that some did suffer injury was almost certain to be reached and would also be reached with a group of healthy people.

However, sometimes we do want to look at many different aspects of a data set. For example, we might have measured 100 different patient characteristics in a clinical trial comparing two groups of patients, and we want to know if they differ on any. (Indeed, if you are going to the considerable expense of setting up such a trial, it makes sense to measure lots of things at the time.)

Fortunately, there are tools for tackling p-hacking and reducing the risk of producing spurious and nonreplicable results when carrying out multiple tests. The earliest approach, the *Bonferroni correction*, dates to the 1930s. This is a way of adjusting the p-value of each individual test to allow for the number of tests conducted. In particular, if we control each of the 100 tests so that each has a p-value of 0.1 percent (a 1 in 1,000 chance of incorrectly rejecting a true hypothesis) instead of 2 percent, then the chance of *at least* one showing significance if they are all true is just 10 percent, rather than 87 percent. That is, if all 100 hypotheses are true, there's now just a 10 percent chance that at least one of the 100 will be rejected as false. That's much more acceptable.

In the past two or three decades some substantially more powerful tools for controlling the problem of multiple testing have been developed. Many of these involve extensions and elaborations of the Bonferroni method, for example, controlling the sequence in which the tests are conducted. But a major advance came with the work of Yoav Benjamini and Yosef Hochberg. They shifted attention from the p-value (the chance of drawing an incorrect conclusion if the hypothesis is true) to the "false discovery rate." The false discovery rate is the expected proportion of errors amongst those hypotheses flagged as false. Arguably, this is the more relevant measure: it tells us the chance we will make a mistake when we conclude that a hypothesis is false.

Apart from p-hacking, another and indeed a more basic reason for failure to replicate results is potential differences in the experimental conditions. Accounts of experiments given in the scientific literature are necessarily brief—the cost of pages in academic journals ensured that was so in the past, though it is of less concern in the age of the World Wide Web. This means that papers rarely describe the exact procedures of any study in great detail. Again, since research is at the boundary of knowledge, slight changes to an experiment can have a big effect on the results.

Yet another cause of mistaken results is the pernicious practice called "HARKing," or *H*ypothesizing *A*fter the *R*esult is *K*nown.

HARKing occurs when a hypothesis is tested using the data which suggested it in the first place. If you look at a data set and spot something in it which leads to an idea, it is unfair then to use the same data set to test that idea to see if it is true. Obviously, the chance that those exact same data will cast suspicion on the hypothesis is pretty small! For example, if I observe that the average weight of 1,000 grains of sand drawn from one beach is greater than the average weight of 1,000 grains drawn from another beach, I can certainly conjecture that in general the first beach has heavier grains than the second, but I can hardly use the weights of those same 2,000 grains to test the hypothesis, since they are certain to support it. The idea must be tested by looking at another, independent data set. We need to find some data we have not previously seen—some previously dark data.

Note that it is perfectly acceptable to sift through, examine, and analyze data to see if they have interesting features. Such "exploratory" investigation is an important (indeed, a fundamental) way of generating new hypotheses and of coming up with new ideas and spotting previously unrecognized phenomena. But those same data cannot be used again to see if those ideas are true.

HARKing can be alleviated by requiring researchers to state their hypotheses before they collect any data. Some scientific journals are considering moving in this direction, guaranteeing publication of a paper regardless of how the results come out, provided the hypothesis is stated beforehand, and the design and methodology meet a rigorous standard.

Hiding the Facts

As we have seen, science is essentially a self-correcting process. The bottom line is always validation by comparing predictions with data, with the consequence that a proposed theory which turns out not to fit the facts will sooner or later be rejected or modified. The flip side is that theories which eventually turn out to be incorrect will often seem to be supported by the data for a while, before the self-correcting nature of science swings things back toward a more accurate portrayal of the truth.

We have already noted that one obvious way this can happen is when there's something wrong with the initial data. Perhaps the observations were insufficiently precise to reject a theory, perhaps they simply had distortions and errors, or perhaps they were incomplete in some way. We have seen plenty of examples in this book, all illustrating the way shadows can be cast over data. But sometimes these distortions and errors might be the work of a deliberate hand—the data might be fraudulent.

We explored financial and other kinds of fraud in earlier chapters. In a sense it is hardly surprising that fraud occurs in areas like finance: the rewards are obvious. But scientific research is not generally a path to great wealth. Indeed, the popular image of science is that is it pursued by disinterested researchers (in white coats, probably!) who are concerned solely with the truth, uninfluenced by earthly matters. Unfortunately, this image does not reflect reality. Scientists, after all, are human, and so are subject to the same drives and motivations as everyone else. Money, power, respect, and peer regard are just as important to scientists as to people in other professions or occupations. Scientists are just as likely to be driven by greed, hubris, and envy as are others.

But there is another key difference between finance and science: it is possible that fraudulent financial transactions might never be discovered, but incorrect scientific assertions will eventually be refuted. The self-correcting nature of science sees to that. So what would be the point of publishing false claims which would ultimately be detected?

One possibility is that the theory actually might not fail. People do make lucky guesses—though guessing is not a sound strategy for a successful scientific career. Another is that the theory might not fail during the perpetrator's lifetime, or conceivably even for centuries. This could tempt someone who is convinced his or her theory is right to adjust or even invent data to make it better match their theory, hoping that no one will spot the deception. Some of the greatest of scientists are alleged to have been guilty of dubious practices with their data, including Robert Millikan, Louis Pasteur, John Dalton, Gregor Mendel, and even Galileo and Newton. We shall look further at the case of Millikan, because there are recorded data on which to judge the allegations in his case—his notebooks have been preserved.

In the other cases, however, the data are spotty at best, so that detection depends on careful statistical forensic work.

One of the most interesting points here is that the people I listed in the previous paragraph are indeed regarded as among "the greatest of scientists," and this is in part because posterity and replication supported their conclusions. Had things gone the other way, they might have been regarded as unreliable or worse and consigned to the dustbin of history. There seems to be an element of unfairness in this!

Even if a theory built on fraudulent data turns out to be incorrect, the fraud might not be detected. The nature of science means there are plenty of failed theories which initially seemed to match the data but which later began to show cracks as replications failed to reproduce the results. But unless there is reason to go back and reexamine past data (assuming they are still available), the failure of the theory will generally just be dismissed as data measurement inaccuracy, chance variation, or some other shortcoming, rather than dishonesty.

Occasionally, however, frauds *are* detected, even long after the event, and then a glittering career can crumble. A common pattern seems to be for someone to begin with a small deception and then, encouraged by its success, to make larger ones. These stack up, until eventually one is spotted, stimulating closer—and retrospective—examination of data and experiments. And before long a whole edifice collapses.

Sir Cyril Burt was a very eminent psychologist: he was the first non-American recipient of the prestigious Edward Lee Thorndike Award from the American Psychological Association, in 1968. But suspicions were aroused about his work on inheritability of intelligence shortly after he died in 1971, when Leon Kamin noted that some of Burt's correlation coefficients (indicating how closely two variables take similar values) from different experiments were the same to three decimal places. The chance of such a precise match is tiny. Several authorities agreed that Burt had committed fraud, although others noted that various researchers had obtained similar correlation values. Arthur Jensen claimed that "no one with any statistical sophistication, and Burt had plenty, would report exactly the same correlation, 0.77, three times in succession if he were trying to fake the data."[21] This is an interesting argument. It suggests that anyone wanting to commit scientific fraud should make it obvious, since no one will believe you could be so foolish. I'm not sure that it's a sound strategy. Note also that it relies on destruction of the evidence: all of Burt's notes had been burned, so no one could check the coefficients or see if the data from which they were purported to arise actually existed.

The initial steps might be of a fairly unobjectionable kind. All scientists have to make subjective decisions at some level about what data are acceptable and what should be rejected. After all, if you discovered that someone had kept his winter overcoat on when he was weighed, or her shoes on when her height was measured, you might feel justified in not including those data in your analysis. But what if the scale broke down soon afterward, leading you to doubt the accuracy of its earlier measurements; or you could not remember if you had instructed subjects to remove their shoes when their height was measured? Do these suspicions mean you should reject these cases? Indeed, it is possible that different researchers might make different decisions.

The possibility of fraud in science is not a new one. In chapter 6, section 3, of his classic 1830 book, *Reflections on the Decline of Science in England, and on Some of its Causes*, Charles Babbage, "the Father of the Computer," said: "Scientific inquiries are more exposed than most others to the inroads of pretenders; and I feel that I shall deserve the thanks of all who really value truth, by stating some of the methods of deceiving practised by unworthy claimants for its honours, whilst the mere circumstance of their arts being known may deter future offenders. . . . There are several species of impositions that have been practised in science, which are but little known, except to the initiated, and which it may perhaps be possible to render quite intelligible to ordinary understandings. These may be classed under the heads of hoaxing, forging, trimming, and cooking."[22] These all represent different ways of making data dark, so let us examine each of them in more detail.

Hoaxing

Hoaxing is the fabrication of data (*DD-Type 14: Fabricated or Synthetic Data*) (or, as we shall see, even actual objects such as fossils, bones, or even complete animals) to give the impression that something exists which really does not, but with the intention that the truth should be discovered eventually, to the embarrassment of those fooled. It is a form of scientific practical joke.

Scientific practical jokes often seem to be perpetrated with the aim of subjecting a pompous target to ridicule. For example, Johann Bartholomeus Adam Beringer, dean of the faculty of medicine at the University of Würzburg in the early eighteenth century, collected fossils. When some extraordinary examples were brought to him, some apparently showing animals, insects, or plants, others with images of stars and planets, and yet later ones engraved with letters spelling out the name Jehovah, he was very impressed, even going to the extent of publishing a book about them. He was convinced the chisel marks which could be seen in some of the stones were stronger evidence that God had created them. Talk about confirmation bias!

At some point the perpetrators, who turned out to be Beringer's university colleagues—J. Ignatz Roderick, professor of geography and mathematics, and Johann Georg von Eckhart, privy counselor and university librarian—decided things had gone too far. They told Beringer it was a hoax, carried out because he was so arrogant. But he didn't believe them. Instead, he suspected they were trying to rob him of the glory of discovery. Only when a stone turned up with his name on it did he accept it. He sued, and Roderick's and Eckhart's careers were ruined.

In a similar example in a lighter vein, two boys tried once to play a trick on Charles Darwin by sticking the head of a beetle, the wings of a butterfly, and the legs of a grasshopper onto the body of a centipede and asked him if he could identify it. Darwin looked carefully at it, and after a pause asked if it hummed when they caught it. On being told that it did, he suggested it was a humbug.

By making up data, the perpetrators of a hoax are concealing the truth, obscuring what real data might look like by replacing them with false data.

A more recent example of a hoax, one which has become a classic, skipped the data altogether and went straight for the published paper. To test the intellectual rigor of the postmodernist journal *Social Text*, physicist Alan Sokal submitted a nonsense article titled "Transgressing the boundaries: Toward a transformative hermeneutics of quantum gravity."[23] *Social Text* is a journal which "covers a broad spectrum of social and cultural phenomena, applying the latest interpretive methods to the world at large."[24] The paper was accepted and published without referees' review, at which point Sokal revealed it was a hoax. Since such hoaxes are a kind of practical joke, they are uncomfortable for the targets (recall Beringer, earlier). However, if they reveal untruths or woolly thinking, they can be useful: "Many researchers in the humanities and social sciences wrote to Sokal . . . to thank him for what he had done."[25]

There's a modern twist to hoaxing of the Sokal kind which deserves mention, arising in the context of so-called pay-to-publish journals. The web has had a very large impact on science publishing, as it has on other walks of life. In the past, researchers or their libraries would take out subscriptions for journals, and these subscriptions formed the basis of the journal business model. However, the advent of the internet and web permitted people to upload their papers to a website free of charge, so a different business model was needed. The system is still in something of a state of flux, but in an important model which has emerged the author pays a fee when the paper is published, and then the paper is free to all potential readers. Unfortunately, a side effect of this process has been that unscrupulous operators have created "journals" which will publish anything for a fee, no matter how nonsensical or fraudulent. There have been many exposures of such publication outlets, in which people following the Sokal model have deliberately submitted nonsensical papers to see if they would be accepted, to ridicule the journals.

A particularly noteworthy such exercise was carried out by John Bohannon, who submitted versions of a paper under the name Ocorrafoo Cobange, from the fictitious Wassee Institute of Medicine, to 304 journals.[26] Of the paper Bohannon writes: "Any reviewer with more than a high-school knowledge of chemistry and the ability to understand a basic data plot should have spotted the paper's shortcomings immediately. Its experiments are so hopelessly flawed that the results are meaningless." But "more than half the journals accepted the paper, failing to notice its fatal flaws."

Another of the many examples is a paper by David Mazières and Eddie Kohler, originally submitted to a conference and later accepted by a journal (which clearly had no refereeing process whatsoever).[27] The paper consisted solely of repeats of the phrase "Get me off your f***ing mailing list." The title of the paper—well you can probably guess.

Such attempts to demonstrate the shallowness of some journals can face significant obstacles. At the time of this writing, assistant professor of philosophy at Portland State University Peter Boghossian is possibly facing the sack after coauthoring a series of papers, seven of which were accepted, in which he lampooned the level of scholarship, aiming to "see if these disciplines that we called 'grievance studies' are compromised by political activism that allows for the laundering of prejudices and opinions into something that gets treated as knowledge." Richard Dawkins and Stephen Pinker have defended him.[28]

Forging

Forging is similar to hoaxing, except that ultimate discovery is not intended. Again, the perpetrators are concealing what real data might look like, turning them dark by replacing them with fake data. Curbstoning—the fabrication of data in surveys and censuses which we met in chapter 2—is an example.

One of the most famous cases of forging was the Piltdown Man. In 1912, lawyer and amateur archaeologist Charles Dawson wrote to his friend Arthur Smith Woodward, who was then Keeper of Geology at the Natural History Museum in London telling him he had found a section of old human skull in gravel beds near Piltdown in East Sussex. Together they continued to investigate, and Dawson unearthed a fragment of a lower jaw, along with some teeth. The result of putting these together with the skull fragment and completing the re-creation using modeling clay suggested they had found an example of the missing evolutionary link between early apes and humans.

The find stimulated great excitement, and not a small amount of controversy. Some people suggested the two bones did not come from the same creature. Zoologist Martin Hinton, convinced it was a trick, went so far as to try to expose the perpetrators. He filed down the tooth of an ape to match one that Woodward had made from modeling clay and planted it in the gravel beds.

As he had planned, the fake tooth was discovered. Unfortunately, however, rather than revealing Dawson as a fraud, the find was regarded as more evidence that he was right. Determined not to give up, Hinton took a leg bone from an extinct species of elephant and carved it into the shape of a cricket bat, which he also buried at Piltdown. But even that failed, as Dawson and Smith Woodward published a scientific paper in *Geological Magazine* on their latest discovery. They wrote: "During the past season we have spent much time in examining the Piltdown gravel, working round the margin of the area previously explored in detail. . . . however, discoveries were few. . . . No human remains were met with; but a large piece of bone evidently worked by man compensates for much disappointment, and proves to be so singular that it is worthy of special description."

It is interesting how people can deceive themselves. In a nice example of confirmation bias, Dawson and Smith Woodward commented: "This bone implement was found about a foot below the surface, in dark vegetable soil. . . . On being washed away, the soil left not the slightest stain on the specimen, which was covered with firmly-adherent pale-yellow sandy clay, closely similar to that of the flint-bearing layer at the bottom of the gravel. The bone, therefore, cannot have lain buried in the soil for any long period, and was almost certainly thrown there by the workmen with the other useless débris when they were digging gravel from the adjacent hole."[29]

The paper, which went into considerable scientific detail about the nature of the artifact and how it might have been made, was followed by a discussion which included the following comments:

"Mr. G. F. Lawrence said that the form of the implement suggest[s] a club."

"Mr. W. Dale said that the tool-marks on the bone were of a character similar to those on an artificially-cut bone in his possession, which was found in the peat, during the dock excavation at Southampton, in association with a hammer-stone of Neolithic age."

"Mr. Reginald Smith said that . . . the discoverers were to be congratulated on providing a new and interesting problem, such as would eventually provoke an ingenious solution." (Indeed!)

"Mr. F. P. Mennell said that . . . it seemed to him very remarkable that so primitive a being as *Eoanthropus* should be capable of making and using any implements at all."

In terms of scientific forgery, the Piltdown Man must surely be one of the more successful, as it was more than 40 years before it was definitively established as being a concoction of orangutan jawbone, chimpanzee teeth, and human skull fragment. The creation is reminiscent of Darwin's humbug. Suspicion fell mostly on Dawson. Indeed, according to archaeologist Miles Russell, some other items in Dawson's private collection were also fakes.[30]

Archaeological and paleontological forgery does not have a direct impact on human well-being. But the forgery of researcher John Darsee might have. Darsee served as chief medical resident at Grady Memorial Hospital, the largest hospital in the state of Georgia, and then at Harvard, where he was offered a faculty position in 1981. But that same year an investigation into his data was prompted when suspicions were aroused by some of

Darsee's colleagues about the accuracy of his laboratory results. After extensive inquiry, the National Institutes of Health concluded that he had fabricated data from experiments he had not actually carried out. So much easier simply to make up the numbers instead of going to all that frustrating and time-consuming effort of doing the real work!

Unfortunately, there are many other such examples. Cardiologist Bob Slutsky received superlative recommendations for his outstanding research—until the cracks began to appear. Eventually, a University of California San Diego committee concluded that Slutsky had distorted and fabricated data in various ways.[31] Medical researcher Jon Subdø published papers on oncology in leading medical journals—until it was revealed that the data from 900 patients described in a paper published in the *Lancet* were entirely made up. The Office of Research Integrity (ORI) in the United States concluded that cancer researcher Dr. Anil Potti engaged in research misconduct by including false data (including a claim that 6 out of 33 patients responded positively to the drug dasatinib when only 4 patients were enrolled, and none responded).

In a case reported in 2017, which was unusual because of the numbers of people involved, the Chinese Ministry of Science and Technology found 486 researchers guilty of misconduct.[32] Their use of dark data was not one of creating or modifying the raw data arising from the experiments but, rather, one of paying referees to comment favorably on their submissions or nominating imaginary referees when the journal editor sought opinions on the merits of the paper.

And so it goes on. If you wish to explore further, the ORI monitors public health service research integrity in America and lists case summaries.[33] And forgery does not arise merely in medical research. Physicist Jan Hendrik Schön reported a single set of data several times as having arisen from different experiments, and Dutch social psychologist Diederik Stapel invented data for many studies—causing 58 of his papers to be retracted.

A belief underlying scientific forgeries is that it's easier and cheaper to invent data which match your pet theory than to collect data (which may not actually support the theory anyway). However, it turns out that inventing data which look realistic is not straightforward.

While life would be so much simpler if all measurements in an experiment gave identical results, all real data have random aspects. Very fine measurement of mass or charge or pressure in a physics experiment will generally give a random scatter of values owing to fluctuations in the background conditions, though it is hoped they will be distributed around the true value of whatever is being quantified. Measuring the height of people in a population leads to a distribution of values because the heights differ. The numbers and weights of seeds produced by plants of a particular kind are not the same for each plant for the same reason. Thus, if we want to make up data which look real, we need to make up this randomness as well.

However, people are simply not very good at making up data which look convincingly random, that is, do not have intrinsic patterns. For example, if you ask people to generate a string of digits at random (for example, 26217833383774811256, . . .), they tend to produce too few groups of the same number (like the 333, 77, and 11 in the example), produce ascending or descending runs of digits too often (e.g., 654 and 4567), repeat sequences of digits, and produce other kinds of patterns too frequently. In fact, Bernie Madoff, whom we met in chapter 1, produced too many pairs of the digits 8 and 6 (both in his financial statements and his fabricated golf scores).

Naturally, everything depends on the degree of sophistication of the fraudster. Someone with deep statistical knowledge will be aware of many of the different kinds of ways fake data can depart from real data and will endeavor to match those aspects. Alternatively, they might copy data from somewhere else or even, with a bit more sophistication, copy the data and also add in small random perturbations. All this makes me wonder if at some point it might have been easier to do the actual experiment rather than go to that much effort to fabricate convincing fake data!

Trimming

Trimming is the adjustment of data to make them fit the theory better. Babbage described it as "clipping off little bits here and there from those observations which differ most in excess from the mean, and in sticking them on to those which are too small." If this is done strategically it can, for example, leave the average value unchanged while giving the impression that the range of values, and hence the uncertainty of the measurements, is smaller than it really is.

There are actually some sound statistical techniques that do something like this and which are advocated in some circumstances to limit the undue influence that unusually high or low (and possibly spurious) values can have on the findings. One such method, *Winsorizing* (after Charles P. Winsor), replaces extreme observations with values a certain distance from the mean. For example, values beyond two standard deviations from the mean might be regarded as unreliable and be replaced with values at two standard deviations. The average of the resulting data is less variable than that of the raw data—but we must be aware that the data have been modified. Failing to report what you had done to the data would mean you were disguising the truth. And note that this technical approach does not go as far as sticking the bits of data you cut off onto the other values!

In a dramatic version of Babbage's trimming with intent to deceive, segments of data are moved or copied wholesale from one part of a large data set. As with simply making up the data, this can save so much effort! I have seen this done with numerical data in alleged fraud cases I have examined, but it seems to me to occur most often with images and photographs, in which an image is described as being something that it is not.

I have also seen what is essentially trimming occur at a higher level. Papers submitted to quality scientific journals undergo a refereeing process in which the papers are sent out to several other researchers to comment on—to say if they think the studies are accurate, well conducted, and of sufficient importance to merit publication. If a referee points out that there appears to be something wrong with the study as described in a paper, authors have occasionally been known to amend it to make it ambiguous so that other referees (and readers) will not be able to spot any errors. And then they submit it to another journal.

For example, the validity of a statistical test or modeling procedure might be based on assumptions which seem dubious on examination of the data, potentially invalidating the conclusions. In one case I came across, the mean and median of a data sample were both reported, and their relative values raised the worry that the distribution was skewed, which would invalidate the statistical analysis conducted later in the paper. When I raised this concern in my referee's report, instead of undertaking some other analysis which could cope with this situation (possibly changing their conclusions), the authors simply deleted mention of the median before submitting the paper elsewhere. Unfortunately for them, the editors of the new journal chose to send it to the same referee!

Cooking

Cooking aims to make data look more accurate and trustworthy than they really are by taking many observations and selecting only those which match the theory most closely. Here's Babbage: "If a hundred observations are made, the cook must be very unlucky if he cannot pick out fifteen or twenty which will do for serving up." The strategy is similar to p-hacking.

One of the most famous alleged cases of cooking was that perpetrated by Nobel laureate Robert Millikan. However, closer investigation suggests that all was not as it seemed: there were certainly dark data, but not due to cooking.

Robert Millikan won the Nobel Prize for Physics in 1923, in part for his measurements of the charge of the electron. Initially working with PhD students but later by himself, Millikan carried out a number of experiments involving balancing the rate of fall of charged droplets of water and oil against the force exerted by an electric field. By measuring the terminal velocity of a droplet, he determined when the force of gravity was balanced by the viscosity of the air, and this allowed him to determine the size of the droplet. Further measurements of speed when the electric field was turned on permitted Millikan to determine the charge on the droplet. Through many such experiments he was able to determine the smallest possible electric charge—the charge of the electron.

For us, the key is that in his 1911 *Physical Review* paper, Millikan said: "These are the only drops of this size which were observed with the present arrangement so that they represent the regular run of observations rather than a selected group. . . . *It is to be remarked, too, that this is not a selected group of drops but represents all of the drops experimented upon during 60 consecutive days,* during which time the apparatus was taken down several times and set up anew."[34] (italics in the original) That statement seems to be very clear: it's telling us that there is no possibility of distortion arising from selection of data, either conscious or subconscious—no *DD-Type 3: Choosing Just Some Cases,* and no making data dark.

And that would be that but for the fact that examination of his notebooks reveals that these were in fact *not* all the data. (If you are interested in seeing a facsimile of his notebook for yourself, it is available at *http://caltechln.library.caltech.edu/8/*). Whereas the paper reported the results of measurements on 58 drops, his notebooks show 175 measurements. This begins to look like a clear case of cooking—possibly even of fraud. At least it seemed that way to William Broad and Nicholas Wade, as described in their book *Betrayers of the Truth: Fraud and Deceit in the Halls of Science,* the title of which says it all.[35]

However, deeper examination by physicist David Goodstein suggests that there's more to this manipulation of data than meets the eye. Of the three factors influencing the movement of the oil drops, two of them—namely, gravity and the electric field—were well understood. But the effect of viscosity on objects as small as the drops Millikan was studying was less so. To allow for this, and so that he could be confident in his results, Millikan had to conduct experiments which allowed him to refine and polish his measurement procedure. The measurements conducted during this phase were not included (even if they accorded with his ideas—Goodstein cites Millikan saying of one such excluded measurement, "This is exactly right, the best one I ever had!!!").[36]

And then there were other reasons for dropping some of the measurements. These included measurements on drops which were very small, so they were heavily subject to Brownian motion; and measurements on drops which were too large, so they fell too rapidly for accurate measurement, and so Millikan felt they should not be included. As we saw earlier, this is the sort of decision all researchers (have to) make. If someone bumps the lab table just as a sensitive measurement is being made, you will probably exclude the result. If you slip as you are adding a chemical to a mixture, you may want to skip that result. As with all aspects of life, there are shades of penumbra between perfect light and complete darkness.

Babbage listed hoaxing, forging, trimming, and cooking, but there are other kinds of scientific misconduct, such as plagiarism, which is concealing the true origin of a piece of work, hiding the origin as dark data, and passing the work off as your own. Often, this takes the form of word-for-word copying of text, or even simply changing the title and author on a paper and resubmitting it! Software tools which match submitted papers against the published literature have been developed to detect this sort of thing, so that copying others' work is now a more risky strategy than in the past.

The reputation of institutions can obviously be damaged by cases of scientific misconduct. This means they might wish to play them down, or even conceal them. But hiding misconduct which later becomes public can be even more damaging. Institutions usually set up independent panels tasked with investigating such allegations. I have served on several.

Retraction

We have seen that the standard approach to self-correction in science is replication—carrying out more studies comparing the theory with data. But there are also other mechanisms. If an error is found in a published paper the authors and the journal editors may retract it, acknowledging that it does not show what it claims to show. That may not necessarily mean the paper is wrong but perhaps merely that it does not adequately demonstrate its conclusions. Papers will also be retracted if fraud or misrepresentation has been detected.

A recent investigation by R. Grant Steen and colleagues on publications indexed by PubMed (a database of publications in the life sciences and biomedicine) showed that the rate of retraction "has risen sharply in recent years" and, alarmingly, that "there has been an estimated 10-fold increase in retractions for scientific fraud . . . since 1975."[37] If that sounds like a terrible indictment of science and scientists, let's put it into perspective. There has also been a very substantial increase in the number of papers published in recent decades, with 21.2 million journal articles indexed in PubMed between 1973 and 2011, of which 890 were retracted for fraud. That's just 1 for every 23,799 published articles, which does not look so bad. Steen et al. commented that "from 1973 to 2011 inclusive, the change in publication rate was greater . . . than the change in retraction rate for fraud . . . or error" but added there were periods within this interval when it appeared that the rate of retractions was accelerating. One of the complications is that retraction must necessarily be retrospective, and editors might be looking further back in time and retracting papers published in the more distant past. Readers interested in what sort of papers get retracted, and the reasons for the retractions, might look at *http://retractionwatch.com*. But we have to recognize that retractions probably represent the tip of the iceberg of poor-quality research publications. Doubtless there are many others which deserve retraction but which are missed.

Incidentally, there's a lesson here for would-be fraudsters. (I hesitated about drawing attention to this lesson, for obvious reasons, but decided that it was more honest to give the big picture—and, in the spirit of this book, not hide the information.) By definition, fewer researchers will read obscure journals, and since fraud is less likely to be detected the smaller the readership, it follows that dishonest papers published in obscure, low-readership journals are less likely to be detected. Of course, while this helps an author's publication rate, it does not help the citation rate, which is the number of times other authors acknowledge the impact of the paper on their own work.

In chapter 3 we saw that the dramatic increase in diagnoses of autism in the United States since year 2000 was at least partly due to increased awareness of the condition. The same sort of phenomenon might be a partial explanation underlying the increased retraction rates of scientific papers: perhaps it is partly due to increased awareness and vigilance on the part of editors, referees, and readers in detecting fraud and error. Indeed, the same is likely to be true for scientific fraud in general, where an apparent increased rate might simply be attributable to the fact that people are more alert for such things, so that more incidents get spotted and reported. Awareness of the potential danger of fraud has certainly recently been raised by some of the high-profile cases mentioned above, contributing to the impression that there's a lot of it about.[38]

Provenance and Trustworthiness: Who Told You That?

A particular genre of dark data has hit the news over the past couple of years, that is, so-called false facts or fake news (*DD-Type 14: Fabricated and Synthetic Data*). Wikipedia defines fake news as "a type of yellow journalism or propaganda that consists of deliberate misinformation or hoaxes." *Yellow journalism* is a term coined at the end of the nineteenth century during the newspaper wars between William Randolph Hearst and Joseph Pulitzer II, describing exaggerated and sensational stories. It derives from a comic character, "the yellow kid" who wore a yellow nightshirt and appeared in Pulitzer's *New World* newspaper. Hearst responded by creating another "yellow kid" for his *New York Journal*. Fake news might not be in numerical form, but they still might be classed as a type of dark data—something you think you know which isn't so. And to the extent that it is perpetrated deliberately, it is a kind of fraud.

The problem of distinguishing the truth from the lies is one which has perplexed humanity from the dawn of time. And it's not one which allows for a strict answer. But there is a helpful strategy in the realm of data, namely, simply to insist on being told where the data come from: insist on being told who collected them or who reported them. Or, as I wrote in an article on this topic, demand an answer to the question, *Who told you that?* when you are presented with data.[39] Insist on being told the *provenance* of data. If the information is not forthcoming, then you can draw your own conclusions about the likely veracity of the information. (Unless, I suppose, there's a question of risk to the source, as can happen in some contexts.) In practical terms this means that all newspapers and websites, all journalists and politicians, should say where they got their information. That allows you to check it—you may not want to or have the time to, but at least you are now able to. While this strategy will not overcome all the challenges (that would be expecting too much), and while it will not prevent someone from deliberately selecting information to support his or her position, it will help.

The word *transparency* is often put forward as a partial answer to dark data, at least that kind which arise in fraud and deception. The idea is that it's easy to see what's going on if it's made public. People can then see for themselves what's happening, and fraud becomes difficult. This is the "shine a light" principle which we have seen before. Western democracies place considerable emphasis on such openness at various levels, and governments are encouraged to publish details of their transactions. For example, the UK's Local Government Transparency Code says: "Transparency is the foundation of local accountability and the key that gives people the tools and information they need to enable them to play a bigger role in society. The availability of data can also open new markets for local business, the voluntary and community sectors and social enterprises to run services or manage public assets. . . . The Government believes that in principle all data held and managed by local authorities should be made available to local people unless there are specific sensitivities."[40] It further says: "This Code ensures local people can now see and access data covering . . . how money is spent . . . use of assets . . . decision making . . . issues important to local people." For example, details of expenditure items over £500 in value for the Tameside borough of Greater Manchester for each quarter are publicly available.[41] These details include supplier, department, description of goods/services, amount, date, and others.

At the individual level, however, things seem to be working in the opposite direction—to protecting privacy—or as we might put it, to promoting the secrecy or darkness of individual data. Thus, on 25 May 2018, the EU's General Data Protection Regulation (GDPR) mentioned in chapter 2 came into force. This imposes obligations on organizations which store and use personal data, and gives individuals increased rights over their data and how those data may be used. Personal data are those aspects of data relating to a living individual who can be identified from the data. The GDPR requires organizations to explain why they are collecting and using the data, and requires unambiguous and freely given consent (or other justifications, such as a legal requirement, or to save someone's life). Individuals have rights of access to the data, and the power to require it to be corrected, erased, and transferred to another data controller. It is also worth noting that these sorts of legal requirements do impose considerable bureaucratic drag on enterprises which process large amounts of personal data.

Incidentally, I've interpreted the word "transparent" in the way it is typically used when people talk about data transparency: meaning that people have access to data. But there's another way to look at it. Something is transparent if you can see right through it, perhaps without even noticing it's there. Windows and eyeglass lenses are transparent. And the painful fact is that many of the most effective frauds and con tricks are based on this notion: things are happening without your seeing them, and everything looks fine—until you notice a discrepancy, and then it all comes tumbling down. "Transparent" in this sense has an uncomfortable similarity to "dark" as far as data are concerned.

This chapter has delved further into how to recognize and overcome the challenges of dark data, in particular in the context of scientific investigations. We looked at the notion of comparing theories with data, of errors made because of lack of data, of fraudulent misuse of data and of fabricating artificial data, at drawing attention only to the largest value in a collection, at the problem of very extensive searches throwing up anomalies, and at the suggestion that "most scientific discoveries are wrong." We also looked at the general principle of understanding the provenance of data—of insisting on an answer to the question, Who told you that?

This first part of the book has explored the various ways in which dark data can cause problems. In Part II we look at how to detect dark data, make allowance for them, and beyond that, how we can actually take advantage of dark data.

ILLUMINATING AND USING DARK DATA

Chapter 8

DEALING WITH DARK DATA

Shining a Light

Hope!

We've seen that dark data can occur for many reasons and that while we might know that the data we see are potentially misleading, we might be entirely unaware that we are not seeing everything. We've seen also that the consequences of such ignorance can be serious—in financial terms and also possibly in terms of loss of life. The picture is not a good one!

So what can we do about it? This chapter looks at ways in which we can peer into the shadows to discern what's hidden there, and ways in which we might be able to mitigate the problems even if we cannot determine exactly what has gone amiss. It outlines ideas, tools, methods, and strategies which have been developed to lead us to the right answers, even if we are enveloped by a cloud of ignorance. The bulk of the chapter discusses situations in which data are missing (e.g., *DD-Type 1: Data We Know Are Missing, DD-Type 2: Data We Don't Know Are Missing, DD-Type 3: Choosing Just Some Cases, DD-Type 4: Self-Selection,* and others), and then at the end of the chapter we briefly turn to discuss data we can see but which are potentially misleading (e.g., *DD-Type 10: Measurement Error and Uncertainty, DD-Type 9: Summaries of Data, DD-Type 7: Changes with Time*). Regardless of the cause of the problem, however, a key plank of the solution is *vigilance*: an awareness of what can go wrong. This is perhaps especially important in situations in which the data themselves cannot give you a hint that something untoward has happened (e.g., *DD-Type 15: Extrapolating beyond Your Data, DD-Type 12: Information Asymmetry, DD-Type 8: Definitions of Data*). The hope is that the many examples in this book, along with the list of *DD-Types*, will help you maintain such a state of vigilance, so you will know at least some of the things to keep an eye open for.

But before we dig into the details, there's a very important and fundamental point to emphasize, namely, *that it's not ideal that data are dark*. This is clearly the case when data are incorrect, but it is also implicit in the very word "missing": it represents the notion that you had hoped for more data, but that something has gone awry. While the following methods might be able to alleviate the problems arising from incorrect and incomplete data, it would be better still if the data were correct and complete. This means that *every effort should be made, both when designing the data collection strategy and when actually collecting the data, to try to avoid errors and incompleteness.*

But what if that is not possible?

Linking Observed and Missing Data

If our strategy to collect a full set of data has failed, then the key to coping with dark data is gaining an understanding of why the data are missing. In particular, we need to explore the relationship between the data, observed or not, and whether an item is missing. With luck this will give us some idea of what sort of values the missing items would have had, and that in turn will allow us to compensate for them.

The starting point for doing this is a very useful classification described by the American statistician Donald Rubin in the 1970s.[1] This classification distinguishes three different types of relationships between observed and missing data. We begin with an example.

Body mass index, or BMI, is a standardized measure of tissue mass in the human body. It is used to categorize people as underweight, normal weight, overweight, or obese and is defined as a person's weight in kilograms divided by the square of their height in meters. Persons are overweight if they have a BMI of greater than or equal to 25, and they are obese if they have a BMI of greater than or equal to 30. Evidence suggests that, compared with people of a healthy weight, obese people are at increased risk of Type 2 diabetes, coronary heart disease, stroke, osteoarthritis, some forms of cancer, depression, and a host of other illnesses. For such reasons, there is considerable interest in weight-loss diets.

In a study of one such diet, observations were taken at weekly intervals over a six-month period, to see how things had improved between the start and the end of the six months. The observations included weight, skin-fold thickness, and BMI, but here we shall focus solely on BMI.

Unfortunately, not everyone remained in the study over the entire six-month period, so there was no final measurement for people who dropped out. The question is, Can we ignore them and their dark data, and simply analyze the data from those for whom we have both start and finish values? We saw some of the problems of dropouts in chapter 2, and given that you have gotten this far in the book, you will doubtless realize the answer is likely to be no, we cannot just ignore the dropouts, but let's explore why in more detail.

In the weight-loss study, some of the dieters were more likely to drop out because they failed to stick to the program and felt too embarrassed to return to the group. A second group, especially those who were not dramatically overweight at the start, found they were not losing much weight and began to lose motivation. So some of those, too, dropped out. And a third group dropped out for reasons not associated with weight loss—some moved because of a job change, and others were too busy to attend the clinic, for example.

For the first of these three groups there is clearly a relationship between the probability of dropping out and the BMIs which would have been recorded had they remained in the study. That they did not stick to the program meant they were likely not to have lost weight as rapidly as they should—or perhaps they might even have gained weight. Rubin called situations like this, in which the probability that data were missing was related to the values they *would have had if they had been observed*, "non-ignorably" missing situations (sometimes they are called "informatively" missing). Clearly, this sort of situation is going to be difficult to cope with, because the dark data are likely to be different from the observed data obtained from those who remained in the trial.

For the second type of dropouts, those who were not dramatically overweight at the start and who dropped out because of loss of motivation, there was a relationship between the probability they would drop out and something which was measured—their initial BMI. Although we don't observe the final BMIs of these people, we do know they've dropped out, and we also know their dropping out was related to things we had measured. Rubin called these "missing at random" observations. The point about this kind of "missingness" is that we have indications things are going or might go wrong.

Finally, the third category consists of people whose reason for dropping out bore no relation to the study. For such people neither the measurements before they dropped out nor the measurements which would have been taken had they not dropped out have any relationship to a tendency to drop out. Rubin called such cases "missing completely at random."

For nonstatisticians, Rubin's terminology is probably not very easy to remember, so I shall rename the three types of missing data mechanism:

I shall describe *non-ignorably missing* as ***Unseen Data Dependent***, or **UDD** for short. Here the probability that an observation is missing depends on values which have not been observed. In the example, the probability that a final BMI value is not observed depends on how high it is: people with larger values are less likely to return to have it measured.

I shall describe *missing at random* as ***Seen Data Dependent***, or **SDD** for short. Here the probability that an observation is missing depends on other data which have been observed. In the example, the probability that a final BMI value is not observed depends on the value of the initial BMI, with people with low values more likely to drop out.

I shall describe *missing completely at random* as ***Not Data Dependent***, or **NDD** for short. Here the probability that an observation is missing does not depend on the data, observed or not, in any way at all. In the example, the probability that a final BMI is missing is not related to anything else in the data, either the other observed values or the values which would have been observed had they not been missing.

The merit of Rubin's categorization becomes apparent when we think about what we need to do to adjust for some data being missing. The last type of missing data mechanism is the easiest, so let's begin with that.

In an ideal world, everyone would have been measured at the end of the six months, as well as the start, so the question to be answered is how omitting the results of those who dropped out will distort the conclusions. Now, the NDD dropouts have dropped out for reasons unrelated to the study (***N***DD—that they are missing does ***Not*** depend on the data, regardless of whether it was observed or would have been observed had they not dropped out). There's no reason that these dropouts will differ in any systematic way from those for whom measurements are taken. In fact, it's just as if we had drawn a smaller sample to start with. On average, omitting their unmeasured weights from the analysis will not affect the results, so we can ignore these people. This is a particularly straightforward situation—and probably a rare one. In this case the dark data don't matter.

If only life were always that simple.

Rubin's second class, the SDD case, is more subtle. Whether or not these people drop out depends on their initial BMI score, which was measured (**S**DD—dropout depends on ***Seen*** data). In particular, those with low initial values are more likely to drop out and not record a final value, whereas those with higher initial values are less likely to drop out.

The important thing to note here is that this kind of missingness does not distort the observed relationship between the initial and final BMI values. For any given initial value there might be *fewer* final values, because some are missing, but those that are observed will be properly representative of the distribution of final BMI values for people with that initial value. This means we can estimate the relationship between the initial and final values just using the values which have been observed: it will not give us a misleading idea of the relationship. We can then use this estimated relationship between initial and final values to give, in turn, estimates of final BMI values for any initial value.

Finally, we have Rubin's first class, the UDD cases. These are the really tough ones. Data are missing because of the BMI values they "would have had," which, of course, we do not know (***U**DD—that they are missing is because of **Unseen** data*). They are not missing randomly or because of the magnitude of other observed values. The only way to estimate those is to get information from somewhere else or to make assumptions about why those values are missing.

Here's another example.

Social statistician Cathie Marsh described a data set consisting of a random sample of 200 husband and wife couples drawn from the adult population of Great Britain in 1980.[2] Our aim is to use this sample to estimate the average age of wives in Britain at that time. Examination of Marsh's data shows there are missing values, with some of the wives' ages not being recorded. The question is whether these dark data have an impact on how to analyze the data and if they invalidate any conclusions we might reach. As in the BMI example, the answer depends on why the data are missing.

The unobserved ages of the wives might be NDD data, where the probability of being missing is not related to any of the data values, observed or not.

The unobserved values might be SDD data, where the probability that a wife's age is missing might depend on other data values which we have seen. To keep life simple, in this SDD case let's suppose the wife's decision about whether to divulge her age depends *solely* on the husband's age and not on the value of any other variable. So perhaps, for example, wives of older husbands might be half as likely to divulge their age as wives of younger husbands. We shall assume we always know the value of the husband's age.

Finally, the unobserved values might be UDD, where the probability that a wife's age is missing depends on that very age itself. This is not far-fetched: at least in the past in the Western world there has been a social convention that it is rude to ask a woman her age, and women are often reluctant to divulge it. Here's a passage from "The Match-Maker," a short story by the British writer Saki in the book *The Chronicles of Clovis*, published in 1911:

> *"The crisis came," returned Clovis, "when she suddenly started the theory that late hours were bad for one, and wanted me to be in by one o'clock every night. Imagine that sort of thing for me, who was eighteen on my last birthday."*
>
> *"On your last two birthdays, to be mathematically exact."*
>
> *"Oh, well, that's not my fault. I'm not going to arrive at nineteen as long as my mother remains at thirty-seven. One must have some regard for appearances."*

This social convention might account for the missing values: perhaps the older women were less inclined to answer.

Dealing with the first case, NDD, is again straightforward. Because the missing observations are not related to any actual data values, we can ignore any couples for whom the wife's age is missing and simply estimate the average age of wives in Britain from the other couples. Doing this means that the sample is smaller than the 200 ages we had hoped for, but it doesn't introduce any bias or systematic distortion into our estimate. Of course, if too many values were missing, so that the sample size was dramatically reduced, then any conclusions we reached based on them would be highly uncertain, but that's a different matter.

But what about the SDD situation? The probability that a wife will say how old she is depends on her husband's age, so we could get a distorted sample of wives' ages. We might find, for example, that the ages of older wives, who tend to be married to older husbands, would be underrepresented in the data. If we ignored this possibility, we would underestimate the average age of the wives.

But that also again shows us how to cope with the problem. While it is true that for husbands of any particular age, not all wives might say how old they are, the wives who *do* say are just a random sample of all wives for husbands of that age (we assumed that whether they said or not didn't depend on anything other than husband's age). This means that the average age of those wives who *do* tell can be used as an estimate of the average age of all wives for husbands of that age. This means (again) that we can study the relationship between husbands' ages and wives' ages just using the age pairs we have. And once we have estimated this relationship, we can use it to explore the expected age of all wives, for any age of husband. Now, it's simple to estimate the overall average age of all wives: we simply calculate the average, using the ages of those who did say and the expected ages of those who didn't.

Finally, the data might be UDD. If the probability that a wife's age was missing depended on that very age itself, with (for example) older wives tending not to give their age, then once again we would have a distorted sample of ages. And now we wouldn't be able to ignore the incomplete pairs, as we did in the NDD and SDD cases. After all, for husbands of any particular age the wives who *didn't* give their age would tend to be older than those who *did*, and we would have no information about which was which. Any analysis which ignored the distortion could be seriously misleading. In this case, the UDD situation, we will need to look elsewhere for a solution.

Much of the early research on how to cope with missing data was carried out by economists. This is perhaps to be expected: economics is a particularly tough discipline because people are not passive objects to be measured but react to the measuring situation—perhaps even by refusing to be measured. In particular, they can refuse to answer questions on the basis of the answers they would have given.

The importance of dark data in economics is illustrated by the fact that the American economist James Heckman was awarded the Nobel Memorial Prize in the year 2000 "for his development of theory and methods for analyzing selective samples" carried out in the 1970s. "Selective samples" is another way of saying you don't have all the data—only samples which have been selected from them. Heckman's approach is known as a "two-stage" or even sometimes the "Heckit" method. It tackles SDD data by first building a model for the process which leads to some data being missing and then uses that to adjust an overall model—similar to that used in the Cathie Marsh example. In Heckman's case he was interested in such factors as hours of work and market wages. An example he used which has now become a classic is the wage a woman earns, which is related to other variables but which is missing if she decides not to be employed (this also applies to men!).

In fact we already encountered economic examples, especially of SDD missingness, in chapter 2, when we looked at financial indexes. For example, the Dow Jones Industrial Average is the sum of the individual stock prices of 30 large publicly owned U.S. companies divided by the Dow divisor. But companies come and go—and the constituents of the Dow Jones have changed over 50 times since it was launched in 1896. In particular, companies might be dropped from the index when they have financial difficulties or as the economy changes. This means that the index is not representative of overall company performance but only of those companies which are doing reasonably well. But the signs of deteriorating performance or economic changes precede and inform the decision to drop a company—the data are SDD.

Likewise, companies will drop out of the S&P 500, a weighted average of the stocks of 500 companies with large market capitalization, because of poor performance relative to others. The decision about which companies to expel must be made on the basis of data available prior to the expulsion (assuming it is not retrospective!), so again the data describing the dropped companies, data which are missing from the calculation of the index, can be described as SDD.

As a final financial index example, in chapter 2 we saw that survivor bias affected not only the Dow Jones and S&P 500 but also hedge fund indexes. The Barclay Hedge Fund Index is based on the arithmetic mean of the net returns of the hedge funds listed in the Barclay database. But funds whose performance has deteriorated to the extent they have been closed will not be included there. Once again, though, the deteriorating performance should be apparent in the months leading up to the closure, so these data may also be SDD.

Identifying the Missing Data Mechanism

The NDD, SDD, UDD classification is very useful, since different types of missingness require different types of solution. Of course, that means we need to be able to identify the class of any particular missing data problem—and if we get it wrong, our conclusions might be wrong. In the example concerning wives' ages, we could be misled if we were mistaken in an NDD assumption that the probability a wife's age was missing was independent of her age or her husband's age. Likewise, if we believed the missing data were SDD but were wrong in our assumption that a wife's decision on whether to divulge her age depended solely on her husband's age, then we could again draw incorrect conclusions. But this should not be surprising. *Any analysis makes assumptions about how the data arose, and if those assumptions are wrong, it naturally follows that the conclusions could be wrong.* But it also follows that we want to be as confident as we can in our assumptions, and if possible find ways to test and validate them. Various strategies have been explored to do this.

Perhaps the most basic strategy is to use your expertise about the domain the data describe. For example, if you are working in an area where people are particularly sensitive about the data, then you might suspect the missing data to be UDD—survey questions relating to use of cocaine are probably more likely to be UDD than are questions relating to use of public transport.

In general, other studies of the same topic, or studies from related areas, might shed light on why data are missing. Harvard statistician Xiao-Li Meng used this approach to yield an elegant quantification of the impact of missing data on conclusions.[3] He decomposed a measure of the accuracy of an estimate into parts, one of which was the correlation between whether a value was missing and the size of that value itself. And then he described how indications of the size of this correlation can sometimes be obtained from other data sources, such as similar problems elsewhere.

A more active strategy for investigating why data are missing is to try to collect some of the missing data. We explore this method in detail in the next section.

Sometimes, statistical tests can be used. For example, we could split the husbands into two groups according to whether their wives did or did not give their age. A difference in the shape of the distributions of husbands' ages for these two groups suggests the data may well not be NDD. American statistician Roderick Little, one of the leading experts in how to cope with missing data, has developed a general statistical test to see if missing data on multiple variables are NDD.[4] Statistical tests to see whether the data are SDD have also been developed—but they are sensitive to any assumptions made about the models. That means if your basic model for what is going on in the data is wrong, so might be your conclusions. Again, this should be hardly surprising.

We have seen that identifying the missing data mechanism, and in particular the extent to which data are missing because of the values they *would have had*, is important in avoiding misleading results. Sometimes it is clear that one or other of the missing data types is appropriate, but at other times we might be faced by a mixture of the different types. The three different processes are not mutually exclusive, and just because some of the missing data are NDD does not mean other missing values are not UDD. Nevertheless, if we are able to classify the missing data values in this way, then we are well on the way to tackling the difficulties of dealing with missing data.

With the tripartite NDD, SDD, UDD missing data classification in hand, we can now begin to explore practical methods for coping with dark data. We begin, in the next section, with some simple and widely used—and it has to be said often misleading—methods.

Working with the Data We Have

Identifying the missing data mechanism gives us a powerful handle on how to cope with the problem. But it clearly requires quite a sophisticated level of understanding, so various simpler methods are often adopted. These methods, which are often obvious and straightforward, are widely available in statistical software packages. Unfortunately, "simpler," "obvious," and "straightforward" do not necessarily mean valid. Let's look at some of these approaches and their properties, and see how they relate to the UDD, SDD, NDD classification.

Table 6 shows a small sample illustrating the sort of data collected at the start of studies of weight-loss diets. Entries labeled "NA" mean that the value in that cell was not recorded.

Complete Case Analysis

First, we might use only those rows of the table which are complete; that is, those rows which have observations for all the characteristics. That would make sense if we believed the dark data to be NDD, and for obvious reasons, such an approach is often called *complete case analysis*. However, if you examine Table 6 and imagine attempting this, you will immediately see one drawback which can arise even if the dark data are NDD; namely, *all* the rows in the table are missing at least one value. If we knocked out incomplete rows, we would be left with no data at all!

This might be a particularly extreme (and, I have to admit, artificial) example, but even in less extreme cases such an approach can mean dramatically reducing the sample size. While we might be happy to draw conclusions on the basis of 1,000 records, we might justifiably be uneasy about doing so on the basis

TABLE 6. Data sample from a study on diet

Age (years)	Height (cm)	Weight (kg)	Sex
32	175	NA	Male
NA	170	90	Male
NA	180	NA	Male
39	191	95	NA
53	NA	86	Male
38	NA	90	Female
61	170	75	NA
41	165	NA	Female
NA	158	70	Female
31	160	NA	Female

of just 20 of those records. Even if the data were NDD, so that the 20 fully observed records were properly representative of the underlying population, the variability arising from such a small sample might make us uneasy about the accuracy of any conclusions.

And, of course, if the dark data were not NDD, then even a small reduction in sample size could mean we were left with a distorted data set.

Use All Available Data

A second simple strategy is just to use all the data we have. For example, seven of the rows have age recorded, so we could estimate the average age using just those seven values. Again, that's fine if the records with missing ages don't differ in some way from the others—if the data are NDD. But if the missing values do tend to differ from the others, we can be misled. For example, if the missing age values in Table 6 tend to be the older ages, then this approach would mean we would underestimate the average age.

There's a further complication which can arise from this method. Different records are missing values for different characteristics: some records are missing ages, but different records are missing weights. That means that if we used this approach, the estimates of average age and average weight would be based on different people. If the weight values of the heavier people tend to be missing, and the height values of the shorter people tend to be missing, this approach could give the incorrect impression that the population consisted of tall skinny people. Indeed, this sort of approach can even lead to contradictions. For example, if we are studying relationships between pairs of variables, we might find that the age/weight correlation and the age/height correlation imply a weight/height correlation contradicted by a direct calculation from the data of weights and heights. That would leave us in something of a quandary.

Missing Value Patterns

A third strategy is to group records into classes according to which characteristics are missing. For example, we could analyze those with weight missing separately from those where weight is not missing. In fact, in Table 6 there are five missing data patterns: those cases with just weight missing, those with just age missing, those with both weight and age missing, those with just sex missing, and those with just height missing. Obviously, with a sample size of only 12 cases there are not many records with each missing data pattern (in fact, 3, 2, 1, 2, and 2 respectively), but with larger samples we could analyze the cases within each pattern separately. This sort of approach could be used with any of the three missing data mechanisms, but it might be difficult to pull all the conclusions together into a useful summary.

Moreover, with large data sets, with many measured variables, there could be a lot of missing data patterns!

This sort of approach makes particular sense when the missing values are not observed because they don't exist. For example, as mentioned in chapter 2, it would make sense if "spouse's income" in a survey was missing because the respondent had no spouse. Then, clearly, we would be dealing with two distinct kinds of cases: respondents who had a spouse (and gave a value), and respondents who did not have a spouse. However, it might not make as much sense if "spouse's income" was missing because someone simply refused or forgot to say what it was.

This example also shows the importance of having different codes for different kinds of missing values. NA can cover a lot of sins, and this simple "unknown" categorization might be unhelpful.

Perseverance and Gold Samples

In earlier chapters we saw many situations in which data for large parts of a population were missing. This might be because people had refused to answer questions in a survey, because a screening process had identified them as probably not suffering from the disease, because a data linkage process had poorly matched different databases, or for a variety of other reasons. If those who had not provided data could be identified—perhaps because they satisfied certain criteria or simply because there was a list (such as a sampling frame in a survey) identifying who had and who had not responded—then a very simple possible strategy would be to follow up those (or some of those) with missing data. If done effectively, this can overcome the difficulties arising from any of the missing data mechanisms.

In fact, this is a widely adopted strategy in survey work, often with a concerted effort made to contact nonrespondents. If repeated callbacks are made in an effort to obtain an interview, it can be possible to model the relationship between the characteristics of the respondent and the number of attempts to interview him or her. This relationship can then be used to adjust the results in an effort to allow for those who could not be contacted at all.

The strategy of using additional data can arise in various guises. Here's an example from earlier.

In chapter 2 we briefly looked at dark data in consumer or personal banking—the kind we engage with almost every day. We saw how difficult it was to build models aimed at all potential applicants, because the available data were typically a distorted sample. For example, we never discover what would have been the outcome (default/nondefault on repayments) for unsuccessful applicants who were not given a loan. Within consumer banking the term *reject inference* has sprung up to describe strategies for trying to infer what would have been the outcomes for those people. These are the "rejects" and are contrasted with the "accepts," who were given a loan. (The consumer credit industry was not known in the past for its tact in choice of terms. So we had "rejects," "subprime applicants," and even "lemons.") Reject inference is a special case of a very general basic strategy for coping with missing values called *imputation*, which will be discussed later in this chapter.

There are various reasons we might be interested in inferring these unobserved outcomes. A basic one is to see how well our selection method is performing—is it, for example, rejecting many applicants who would not have defaulted? A second one is to build better models to predict the likely outcome of new applicants. After all, if our models are based on only part of the population—those previously given loans—then they might be quite misleading when applied to the entire population of applicants. This was a problem described in chapter 1.

To tackle the problem of the unknown outcomes of people not given loans, one bank I worked with had obtained what they called a "gold sample." This was a sample of people who *should have been rejected* because they did not meet the bank's previous criteria for a loan. They were thought to have a high probability of defaulting, but the bank nevertheless accepted a (small) random sample of them because of the information that could be obtained by giving them a loan. This strategy allowed the bank to build a better model of the sort of people who were likely to default on loans and to make better decisions about whom to offer loans to in the future.

Unfortunately, it is not always possible to draw another sample to make up for the missing parts of the population, but sometimes we have an idea of what the overall shape of a distribution should look like, perhaps from other similar problems (e.g., the age distribution of the populations of similar countries) or from theoretical arguments (e.g., the shape of the distribution of lifetimes of lightbulbs based on the physics of how lightbulbs work). In such situations, provided we know the selection criteria, we can use the part of the distribution that we do observe to estimate the entire distribution—and hence features of it, like its average value. The next section gives an important illustration of this idea.

Going Beyond the Data: What If You Die First?

We are often interested in *how long* it will be before some particular event occurs. For example, we might want to know how long someone will stay in a job, how long a marriage will last, or how long it will be before an engine breaks down. In surgery, it is sometimes necessary to lower a patient's blood pressure by using a hypotensive drug, but doctors want blood pressure to return to normal as soon as possible after the operation. Consequently, it's important to know how long that will take and whether the time taken to return is related to blood pressure levels during surgery. More generally in healthcare, we might be interested in how long it is likely to be before someone dies, until a disease recurs, or until an organ fails, and so on.

Problems of this type are called problems of *survival analysis*. They have a long history, especially in medicine, and they are closely related to life tables in actuarial work, concerning how long people will live in general, and to methods of reliability analysis in manufacturing, concerning how long objects will last before breaking down.

The difficulty of estimating likely survival times is illustrated by the example of patients with stage III prostate cancer. Stage III means there is evidence the cancer has spread into nearby tissues. To determine which of two treatments is the more effective in prolonging life, patients will be randomly assigned to the two treatments, and the average times for survival in the two groups will be compared. But inevitably some patients will survive for a long time—perhaps decades—and we will not want to wait for decades before knowing which treatment is better. Thus, the study will probably be terminated before all the patients have died. That means we will not know the survival times for those patients who lived beyond the termination date—the data will be missing. Additionally, other patients might die of something else instead of prostate cancer. For these patients also, the data telling us how long they would have survived before dying of cancer are missing. And, as usual, yet others might drop out for reasons unrelated to the study. Their survival times are again dark data.

It is immediately obvious that if we simply ignore the patients for whom we do not see the real survival times, then we could produce grossly misleading conclusions. For example, suppose that one of the treatments was very effective—so that all but a couple of patients receiving this treatment survived beyond the end of the study. Then we would dramatically underestimate the effectiveness of this treatment if we ignored all but those two patients.

However, although we won't know the survival times of patients who lived beyond the time at which the study was terminated, or who died from other causes, or who dropped out for other reasons, we will know the times at which they stopped being measured. These times are called "censored" times, and such censoring means we know that the time interval between when the patients entered the study and when they *would have* died of prostate cancer is *larger* than the length of time they were in the study.

In a very important paper which appeared in the *Journal of the American Statistical Association* in 1958, Edward Kaplan and Paul Meier showed how to estimate the probability that people will survive beyond any given time, allowing for the fact that some survival times will be longer than the observed times.[5] The importance of this paper is illustrated in a report by Georg Dvorsky showing that the Kaplan-Meier paper is the 11th most highly cited scientific paper of all time.[6] Given that there are more than 50 million scientific papers, this is some achievement.

Sometimes we want to go beyond merely estimating the probability that people will survive longer than any given time. For example, we might want to estimate the average survival time.

Now, distributions of survival times are typically positively skewed, meaning that longer times are less common than shorter times, and there might be many short times but only a small handful of very long times. Statisticians describe such a distribution as "having a long tail." Given that the top few values in a positively skewed distribution might be *a great deal* larger than most of the values, not including them in the analysis could have a huge effect on an estimate of the average value. Think of calculating the average wealth of people in the United States leaving out Bill Gates and every other billionaire. The result would be a serious underestimate. In terms of survival times, it could mean omitting those who survived longest, which could obviously dramatically distort any conclusions.

So what can we do about this problem?

The ideal of drawing an extra sample, this time of those who dropped out before dying from prostate cancer, is out of the question. We cannot, for example, follow up a sample of people who died from other causes to see how long they would have lived before dying of prostate cancer had they not died of other causes (the counterfactuals again).

So, instead, we have to resort to other ways of modeling the distribution of those we are unable to follow up. A common strategy is to assume that the overall distribution of survival times follows some familiar shape. This assumption might be based on past experience and observation with other diseases. To illustrate, one common assumption is that the survival times follow an *exponential* distribution. This is a particular family of positively skewed distributions with many small values and just a handful of very large values. The choice of which member of this family to use can be determined using the observed survival times along with the fact that the censored times must be greater than the observed times of dropping out of the study.

While this approach might be reasonable in many cases, we must not forget that we made the assumption that the exponential distribution was appropriate. As always, if the assumption is unrealistic, the conclusions could be wrong.

Survival analysis combines the known survival times of those whose deaths from the cause in question were observed with the known fact that others' survival times were greater than a certain time. If we could have estimated the survival times of these others, we could have simply summarized all the times—the observed ones and the estimated ones. That suggestion leads us on to a very general method of coping with missing values, namely, imputation. This is the subject of the next section.

Going Beyond the Data: Imputation

A natural suggestion for attempting to tackle incomplete data is to complete them by inserting substitutes for any missing values. This strategy is called *imputation*. Since it makes the data complete, once we have imputed missing values we don't have to worry about any holes, and we can go ahead and analyze the data any way we want. For example, once values have been plugged in for the missing ages in Table 6, we can very simply calculate the average age of all 10 people in the sample. However, this approach does sound suspiciously like making up data, and unless we want to be accused of fraud we need to think carefully about how we are doing it. Moreover, it's one thing if the unobserved values are NDD but another if they are SDD or, worse still, UDD. If the missing values are SDD, we will want to make our imputed values depend on aspects of the observed data. If the missing values are UDD, the observed data tell us little about what the imputed values should be—and plugging in inappropriate values could lead to misleading overall results.

One reason that imputing missing values can often simplify analysis is that many statistical methods are based on *balance* and *symmetries* in the data. For example, I advised a manufacturer of injection-molded plastic car-parts who wanted to know what combination of levels of three factors—temperature, pressure, and time in the mold—would yield the highest-quality product. Two levels of temperature, two levels of pressure, and two levels of time were to be tried. (In fact, they explored more than two levels, but I shall take two here to keep things simple and call them high and low for each factor.) With two levels for each factor, there are a total of eight combinations: all three at the high level, the first two high and the third low, and so on. The manufacturer carried out several production runs at each of these eight combinations, and each production run yielded a finished part whose quality could be assessed. In experiments like this, if the same number of parts is produced at each combination of the three factors, then convenient mathematical formulas can be used to give the results. But the analysis is more difficult if the different combinations produce different numbers of car parts. In particular, if the original design aimed at having such a well-balanced number of cases, with the same number in each factor combination, but some values are missing (e.g., a power cut disrupted some runs of the manufacturing process), then things become *unbalanced*. This can make the analysis much more complicated, requiring more elaborate calculations. Clearly, the idea of plugging in replacement values to *rebalance* the data is very attractive.

Imputing values for the missing observations is useful, but it's pretty obvious that if we repeat the exercise with different plugged-in values, we'll get different results (remember the phrase "making up data"). Since the motivation for plugging in values is simply to make the calculations easier and not distort the results, we might try to find plug-in values so that a simple calculation based on the balanced completed data will give the same results as a long and complicated calculation using just the incomplete data.

That's an attractive idea, and in some simple situations it is possible, but it seems to be a bit circular: how do we find those critical plug-in values that won't affect the results without carrying out the longer calculation in the first place? We'll return to this question later, where we will see that attempting to answer it can lead to deep insights into what's going on in the data. First, however, we look at basic approaches to imputation in more depth.

Average Imputation

One common approach to imputation is to replace missing values by the *average* of the recorded values. Thus, we could replace the three unknown values of age in Table 6 with the average of the seven known values. Indeed, this very simple strategy is readily available in many data analysis software packages. But no doubt you have already acquired a certain skepticism regarding simple strategies for coping with dark data and are thinking about what could be wrong with this approach. One possible problem is the one we keep meeting: if the missing values differ systematically in some way from the others, then replacing those missing values with the average of the others could well be misleading. For example, if the three people whose ages are missing are older than the other seven, it would not be a great idea to replace their ages with the average of the other seven. So this solution would be fine if the missing data were NDD but could be problematic otherwise.

Unfortunately, however, there's another problem with imputing using the average. It's usually unlikely in the extreme that all the missing values would actually have been identical if they had been measured. That means that plugging in the same value for all the missing values is making the "completed" data artificially homogeneous. If we did this for age in Table 6, for example, then the variance (a measure of how much the values differ) of the completed sample of age values would probably be smaller than the actual variance would have been if all the ages had been observed.

Last Observation Carried Forward

The values missing in Table 6 don't appear to have any pattern but are missing haphazardly. In contrast, as we have seen, we often find that people drop out as time progresses, so that each record is complete until the dropout time, and then all its later values are missing. We saw a very pronounced example of this in chapter 2, Figure 4.

When this dropout pattern occurs, we can use the particular imputation method called *last observation carried forward* (LOCF), which does what it says: a missing value for a patient is replaced by the most recent preceding recorded value. Note that this essentially assumes that things don't change between the time at which the measurement was taken and the time of the missing value. That's quite a brave assumption ("brave" being a possible euphemism for foolhardy.) You might ask yourself about the wisdom of such an approach, bearing in mind that generally the reason for repeating observations as time progresses is that we think things change over time.

Thus, perhaps unsurprisingly, the LOCF method has come under criticism:

- "If there were a prize for the most inappropriate analytical technique in dementia research, 'last observation carried forward' would be the runaway winner."[7]
- "All analyses using LOCF are of questionable veracity, if not being outright specious (definition: appearing to be true but actually false). . . . LOCF should not be employed in any analyses."[8]
- "Both LOCF and mean value substitution falsely increase the stated precision of the estimates by failing to account for the uncertainty due to the missing data and generally give biased results"[9]
- "The use of LOCF is statistically un-principled, with assumptions that are only occasionally justifiable."[10]

Given those comments, you might hesitate about using LOCF.

Predict from Other Variables

So far, the imputation methods we have considered have been straightforward: use the average of the values which have been observed, or use a previous value from the same object, patient, and so on. However, a more sophisticated strategy would be to model the relationship between the variable with a missing value and other variables, and then predict the missing value using observed values of those other variables. The model could be based on those cases with all the values observed. In fact, we met this idea when we looked at the definition of SDD missingness.

For example, in Table 6 there are four rows with values for both age and weight. The four corresponding points are shown plotted in Figure 6. We could use these to construct a simple statistical model relating age and weight. The straight line in the figure might be a suitable model, capturing the fact that, for these

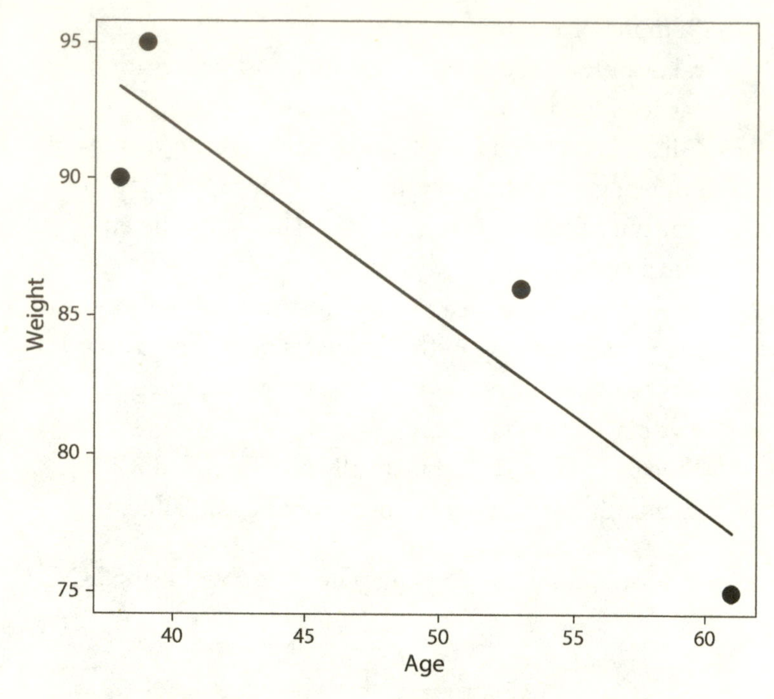

FIGURE 6. Relationship between age and weight for rows in Table 6, in which both are recorded.

data, larger age values are associated with lower weights. (I would not recommend you fit models on samples as small as four points in real life!) Then we could use this model to predict missing weight values when the corresponding age value was recorded. For example, the person described in the eighth row in the Table has an age of 41 but an unknown weight. The line, our model, suggests her weight might be around 91 kilograms.

This is really an extension of the idea of just plugging in the average of the observed values of weights but using a more elaborate statistical model which takes advantage of other information available in the table. Since it uses more information (the values of age and not just of weight), the strategy can yield better results than just using the average. In particular, it is appropriate if the weights are SDD missing, with the probability of their being missing depending only on age. But if the data are UDD missing, the model is problematic. Nonetheless, it turns out that this modeling and prediction strategy contains the seed of a very powerful idea, which we will return to later.

Hot Deck Imputation

Another simple imputation method based on the values of observed values goes under the exciting name *hot deck imputation*. To find a substitute for a missing value in a partial record, it matches the partial record with other records with recorded values in common. Then, it randomly chooses a record from those which match most closely and uses that record's value for the missing item. For example, in Table 6, the first female has an unknown height. Comparing this row with the others, we find two other rows which are similar, in that they are for females with ages of 41 and 31, not too far from the age of 38 of the female with the unknown height. So we randomly pick one to plug her height value into the NA for the 38-year-old. The heights of these females are 165 and 160 centimeters. If we picked the 41-year-old for example, we'd plug in the value 165 centimeters for the missing height of the 38-year-old.

This method derives its name from the days when data were stored on punched cards, and in the past it was used quite extensively. Its attraction is that it's straightforward and doesn't involve any complicated statistics; it simply requires calculating a measure of similarity between the records. However, it does depend on how you define "similarity": Which of the other variables are you going to use in your measure of similarity (all of them?) How do you combine them to yield an overall measure?

And should some of the other variables be regarded as more important and weighted more heavily than others?

Multiple Imputation

We have already seen that one of the apparent problems with the imputation approach is that repeating the exercise using different imputed values will produce different results. In fact, however, we can take advantage of this.

Each completed data set represents a possible configuration of data which could have been observed. A summary statistic calculated from such a completed data set represents a possible value of that statistic which might have been obtained had the data been complete. That means that if we repeat the imputation multiple times, using different imputed values, we will obtain a *distribution* of summary statistic values, one for each possible completed data set. We can estimate various parameters from this distribution, such as the uncertainty or variance of the summary statistic. That is, instead of simply getting a single "best estimate," we will now get a measure of how confident we can be that it takes different possible values.

This strategy of repeating the imputation is called, reasonably enough, *multiple imputation*. It has become a very widely used tool for tackling missing data problems.

Iteration

We have looked at some straightforward methods of substituting replacement values for missing data—using the average of that variable's observed values, for example. We have also looked at more elaborate methods of using other observed variables to lead to an estimate of a missing value. This idea, of using an estimated relationship to predict missing values from observed values on the same record leads to a powerful iterative idea, based on the law of *likelihood*.

For any given set of data, and any proposed statistical model for the mechanism which might have generated them, we can calculate the probability that such a dataset could arise from that model. The law of likelihood then says that of two statistical models we should prefer the one for which the probability of generating those data is the higher. More generally, if we have several or many or even an infinite number of possible explanations for a set of data, the principle says that we should choose the one with the highest probability of having produced the data. The iterative approach is a method for finding that model when there are NDD or SDD missing data.

We begin by plugging in initial substitutes for the missing values (we get these initial values however we want—even random guesses). Then, we use the entire completed dataset—the observed values along with the values which have been plugged in in place of the missing values—to estimate the relationships between variables, using the maximum-likelihood approach. Next, we use those estimated relationships to get new replacements for the missing values. We then use the completed data with these *new* replacement values to give further estimates of the relationship, and so on, round and round. Under certain general conditions, the replacement values change less and less at each step, and the final estimated relationships between the variables are those which give the maximum-likelihood model.

This looping round and round, choosing values for the missing items and then using the completed data to obtain estimates of relationships between variables, which can in turn lead to new substitute values for the missing items, is a powerful idea which has been discovered many times. But in a seminal paper published in 1977, three statisticians, Arthur Dempster, Nan Laird, and Donald Rubin (the last of whom we have already encountered) pulled the ideas together in a remarkable synthesis showing the commonality of the ideas, and setting them in a more abstract form, permitting wide generalization to other situations. They called their method the *EM algorithm*, where EM stands for "Expectation Maximization." This name refers to the two steps in each loop: first the calculation of an "expected" value for each missing item and, second, the estimation of the relationships between the variables using the completed data set. The second step in each loop is called "maximization" because it maximizes the likelihood.

In fact, Dempster, Laird, and Rubin showed that it was not necessary to go so far as to find substitute values for the missing items. All that was necessary was to model the distribution from which the missing values were likely to have come. We encountered this concept when we looked at survival analysis, where we did not try to estimate the survival times of those who lived beyond the end of the observation period but merely used the probability they would live longer than that time.

The EM algorithm has been increasingly strengthened. Once Dempster and his colleagues' abstract description of the basic two-step looping of the algorithm was presented, it became apparent that the fundamental idea cropped up everywhere, and often in all sorts of unexpected ways. Moreover, the algorithm has been extended in a variety of ways—for example, modifications have been developed which require fewer turns around the "expectation-maximization" loop before the substitutes for the missing values (and the relationships between variables) barely change at all, that is, before the process *converges* to the best (in the maximum-likelihood sense) model.

The abstract presentation of the EM algorithm has also led to greater insights. In chapter 1, for example, I mentioned that in general any unknown characteristic of a population can be thought of as missing and hence dark data. This is a very powerful idea. The underlying characteristics might be a single simple value (like the average height of a population), but they could be much more elaborate, involving multiple unobserved postulated variables being related in complex ways. A classic example is the trajectory of a moving object when its precise position is never observed but only values of its position contaminated by measurement error. In many cases the underlying dark data are more than merely unobserved—they are intrinsically *unobservable*. They are hidden or "latent" (and hence *latent variable models*). But that does not mean that light cannot be shed on them, and that is indeed the whole purpose of statistical methods. Statistical tools applied to data generated from the underlying reality tell us something about that reality: the data shine beams of light onto it.

This chapter has explored how we can analyze data and gain a good understanding of the processes which generated them even if there are dark data. We have looked at a range of general methods, from simply working with the available data, through discarding any incomplete records, to a variety of imputation methods in which attempts are made to determine what the missing data would have been had they been measured. We also looked at an important classification describing the structure of missing data—whether they were related to data which *had* been observed. This three-part classification—which I called NDD, SDD, and UDD missingness—led to a deeper understanding of what might be achieved in terms of coping with dark data. In the next chapter we change tack and see how we can actually take advantage of dark data. In particular, we will begin by looking back at some of the ideas we encountered earlier and see how they can be examined from a different perspective which reveals that they rely on dark data. But first, a few general words about data which we *can* see but which are deceptive.

Wrong Number!

Until now, this chapter has focused on missing data. But as we have seen, data can also be dark in other ways—such as *DD-Type 10: Measurement Error and Uncertainty*, *DD-Type 9: Summaries of Data*, and *DD-Type 7: Changes With Time*. Consideration of these leads us to a higher-level perspective on dark data and to three fundamental steps in coping with them, namely, *prevention*, *detection*, and *correction*.

Prevention

Data errors are *prevented* by being aware of what sorts of errors occur and putting systems in place which stop their arising when the data are collected. Of course, "being aware of what sorts of errors occur" may well be because they are mistakes you've previously made or, better still, because you have seen others make the mistakes. (As I once heard someone saying to a manager when he left the organization: "Thank you for giving me so many opportunities to learn from your mistakes.")

If we input data directly into a database, then simple checks can be made as they are entered. For example, if a birth date is required, it is a simple matter for the machine to check that it is a legitimate date. However, sometimes some care is needed. I heard of one case in which a data set had a strange peak of births on 11 November 1911. It turned out that the birth date had to be entered as six digits, in the day/month/year format, and that the programmers were aware that people would sometimes type 00/00/00 to avoid giving a date. If six zeros were entered, the machine rejected the date and required them to try again. And at that point those who were very keen not to give a date would presumably try the next simplest thing they could think of—a sequence of six 1s, yielding an apparent date of 11 November 1911.

Redundancy in the data can be used as a general strategy to prevent errors. This essentially involves entering the data, or at least aspects of them, in more than one way. One common method, especially in clinical trials research, is the *double data entry system*, in which values are copied across (e.g., from data collection forms into the computer) by two different people independently. The chance of their both making the same mistake in the same place is small.

A rather different redundancy strategy involves entering a string of numbers and also entering their sum. The computer then adds up the numbers and compares its result with the input sum. An error in any of the numbers in the string will mean the two sums won't match (except, presumably, in some very rare cases when two errors canceled out). There are some very sophisticated versions of this "check digit" idea.

Detection

The 11 November 1911 example and the use of check digits to prevent errors at the data entry stage verge on error *detection*. Data errors might be detected because they are out of line with other data points or with what is expected. A value of 10ft 5in in a database of human heights would immediately arouse suspicion—no human has ever been that tall. It's possibly a data entry error for 5ft 10in (although we cannot *assume* that that's the case, and we would ideally go back to the data source to check, though this is not always possible).

Errors can also be detected if there are logical inconsistencies. If the claimed number of children in a family does not match the number of ages given in a list of their ages, then there's something amiss. But errors can also be detected from statistical inconsistencies. A 4-foot-tall person weighing 400 pounds might arouse suspicion of a data error, even though some people are 4 feet tall, and some people do weigh 400 pounds.

A more sophisticated example of the statistical detection of oddities occurs in the *Benford distribution*. The first description of this distribution (sometimes called a "law") seems to have been in 1881 by the American astronomer Simon Newcomb. He used books of logarithms in his work. These are tables of numerical values enabling large numbers to be multiplied together very quickly, which were widely used prior to the computer age. Newcomb noticed that the early pages of tables were much more heavily thumbed than later pages. The law was rediscovered nearly 60 years later by the physicist Frank Benford, who made an extensive study showing that the same phenomenon—early values being used more frequently than later values—cropped up with many lists of numbers.

So what is Benford's law?

First, we must define the most significant digit of a number. It's basically the first digit: the most significant digit in 1,965 is 1, and the most significant digit in 6,009,518,432 is 6, for example. Now, in a collection of numbers you might expect the digits 1, 2, 3, . . . , 9 to occur about equally often as the most significant digit. That is, you might expect each of 1, 2, . . . , 9 to occur as the first digit in about one-ninth of numbers. But curiously, in many natural collections of numbers the proportion of times that the digits 1, 2, . . . , 9 occur as the most significant digit is not roughly equal. Instead the 1 occurs about 30 percent of the time, the 2 about 18 percent of the time, and so on down to the 9 occurring about 5 percent of the time. In fact, there is a precise mathematical formula giving the distribution—Benford's distribution.

There are sound mathematical reasons why this strangely counterintuitive phenomenon might occur, but we will not go into them here.[11] For us the important thing to note is that if data depart from this Benford distribution, then it might be worthwhile checking to see if there is something funny going on. In fact, Mark Nigrini, an expert in forensic accounting, has developed tools based on Benford's distribution for detecting fraud in financial and accounting records. And there's a general point here. Tools for detecting data peculiarities arising from errors can also be used to search for peculiarities arising from fraud—where the true numbers are being deliberately concealed. In chapter 6, I mentioned that anti-money-laundering regulations require payments of $10,000 or more to be reported to the regulators. Criminals have tried to overcome this hurdle by splitting their transfers into many sums of just less than this figure. But an excess of numbers beginning with 9 (e.g., $9,999) will show up as a departure from the Benford distribution.

I've spent quite a bit of time working with consumer banks developing tools for detecting potential fraud in credit card accounts. Many of these tools are based on searching for odd-looking data points, values which might be errors but which might indicate deception.

There's one final important point to note about error detection; namely, you can never be sure you have detected *all* the errors. The unfortunate truth is that the presence of errors can (sometimes) be proved, but the absence of errors cannot. As we saw earlier, while data can be wrong in an unlimited number of ways, they can be checked in only a finite number of ways. However, some sort of *Pareto principle* undoubtedly applies, which says that most of the errors will probably be detectable with relatively little effort, but also implies that there's a law of diminishing returns. If you detect 50 percent of the errors with a certain amount of effort, then the same amount of effort again might detect 50 percent of the *remaining* errors, and the same amount again . . . , but you'll never get them all.

Correction

The third strand to coping with dark data, after prevention and detection, is *correction*. Once you've spotted that a number is wrong, you need to decide what it ought to be. How you correct errors—and even if it is possible to do so—will depend on what you know about the values that should really be there and general insight into the sorts of errors that arise. The examples of misplaced decimal points in chapter 4 illustrated how our general knowledge of the data and our experience with errors in the past meant it was often fairly obvious what the true value should be. Likewise, a recorded value that a cyclist was traveling at 150 mph in a table where all the other values lie between 5 and 20 miles per hour suggests that perhaps the true value was 15 miles per hour, based on the context. But we must be cautious about adjusting values too readily—in September 2018, champion cyclist Denise Mueller-Korenek established a land-speed bicycle record of 183.942 miles per hour. Short of going back to the original source and repeating a measurement, we cannot be sure what number should be there, even if we can be sure that the recorded value is wrong.

One final general point about data errors is that the power of the computer has opened up amazing new worlds of data and of understanding. Truly colossal data sets are being collected, stored, and processed thanks to modern computers. These databases lead to tremendous opportunities. But that very computing power also implies a fundamental opacity. Although it enables us to see things in data that we could never discern with the naked eye, the computer necessarily acts as an *intermediary* between us and the data. The computer hides aspects of the data from us.

Chapter 9

BENEFITTING FROM DARK DATA

Reframing the Question

Hiding Data

Dark data might seem to have only a downside. It is certainly true that caution is the key message of this book. But there are ways in which we can use dark data to our advantage, provided we know what we are doing and tread very carefully. That is, there are ways in which we can turn the apparent obscurity implied by dark data against itself to enable us to gain greater understanding, make better predictions, choose more effective courses of action, and even save money. We do this by strategically ignoring parts of the data and by deliberately throwing them into the shadows.

In this chapter we begin to explore this idea, initially by *reframing* some familiar statistical ideas. That is, we look at them in a nonstandard way and view them from the perspective of actively hiding information or data. In fact, the familiar ideas we will initially examine are all ones we have already encountered in earlier chapters. Then, later in the chapter we will look at some novel dark data perspectives on more advanced statistical ideas and methods.

A basic illustration of the nonstandard perspective is given by drawing a sample from a finite population. In chapter 2, in particular, we examined methods of survey sampling, and there we looked at the dark data challenge of nonresponse. However, surveys represent one of the most familiar and straightforward ways of taking advantage of dark data. Surveys, quite correctly, are typically described as using the values of a (random) subset of all the members of a population. But an alternative way of looking at them would be that they are drawing a sample to discard, ignore, or treat as dark. After all, drawing a 10 percent sample of the population to analyze is equivalent to drawing a 90 percent sample to ignore. The general idea is that whenever we are working with a sample of data, we can think of it either as selecting that sample or as discarding the remainder of the population and turning them into dark data.

Note that random selection (or at least "probability sampling") is critical here. Choosing in any other way is likely to lead to the sorts of problems described throughout this book. Using random selection means the missing values are either NDD or SDD, and we saw in chapter 8 that those types of dark data can be coped with.

Hiding Data from Ourselves: Randomized Controlled Trials

Choosing a sample to analyze, and hence also choosing a sample to discard, is the most basic way of taking advantage of dark data. Another important way is via a randomized controlled trial of the kind also discussed in chapter 2. To take a simple example, suppose we wish to determine if a proposed new treatment for some illness is better than the standard treatment. As we have seen, the basic strategy is to randomly assign each patient to one of the two treatments and to compare the averages of the outcomes in each group.

The randomness of the assignment has an intrinsic *fairness*. It removes the allocation from the realm of human choice, making the process opaque and unmanipulable and not subject to either deliberate or unconscious biases. This merit has long been appreciated: as the Bible puts it, "The lot causeth disputes to cease, and it decideth between the mighty" (Proverbs 18:18).

The idea of randomly allocating the patients to treatments is clearly very powerful. It essentially means that we can be pretty confident that any measured differences between groups are due to the treatments and not to some other factor. Or, to put it another way, it *breaks causal links*: it tells us that the *cause* of any difference is unlikely to be preexisting differences between people. By breaking some causal links it means that differences in outcome must be explained by other differences—by the different treatments people received, and not by their age, sex, or other factors.

But perhaps mere random assignment is not sufficient. If researchers can see which treatment patients are assigned to, even if that allocation is made randomly, there might be a temptation to manipulate things. You might feel sorry for the patients receiving the inactive placebo and take more care of them. Or you might more rigorously interpret criteria for withdrawing a patient suffering from adverse side effects if you know that she received one treatment rather than the other.

This danger can be overcome if the identity of the groups to which the patients are assigned is concealed, so that neither the patients nor the treating doctors know which patients receive which treatment. Hiding the group labels in this way is called *blinding*, making them dark data in a very literal sense!

For example, in a comparative trial of two medicines the two drugs could each be assigned a different code, without the doctors knowing which medicine each code corresponded to. If the medicines were then packaged identically, apart from being labeled with the code, the doctors would not know which treatment they were administering and so could not (consciously or subconsciously) act more favorably toward those patients who received a particular medication. Indeed, the same applies to the data analysts, who would be able to see the code indicating to which treatment each patient has been assigned but would not know which actual treatment this meant.

The code identifying which treatment each patient received would be revealed only after the trial had been completed and the data analyzed, and only then would it be possible to see which treatment was the more effective. (The code would always be available to be broken if serious adverse side effects arose.)

What Might Have Been

We have seen that the fundamental idea behind randomized trials is the use of dark data in allocating individuals to treatment groups. This enables us to explore the counterfactual of *what might have happened*. An alternative strategy for investigating what might have happened is *simulation*. In simulation we create a model of a mechanism, system, or process and generate synthetic data from that model to see how the mechanism, system, or process behaves under different circumstances or conditions, and perhaps at different times. The data we generate are not dark data in the sense of values which exist but have not been observed. But they are dark data in the sense of being values which were not observed but could have been observed had things been different—like an unmarried person's spouse's income, or the time of death due to cancer of a terminal cancer patient who died of something else first, or if random measurement error had taken different values.

Simulation is an immensely powerful tool and one which is nowadays used in a vast number of domains, from financial systems, through nuclear weapons policies, to the impact of pollution and human behavior. Indeed, it is such a powerful tool that some philosophers of science have gone so far as to describe it as a new way of doing science. But to introduce things gently, let us first look at a couple of very straightforward examples.

In his book *The Naked Surgeon*, Samer Nashef describes a medical example of simulation. Nashef's study was motivated by the case of Harold Shipman, Britain's most prolific serial killer of all time. Shipman was a family doctor, convicted of killing 15 of his patients but believed to have killed more than 250 over a 25-year period. Nashef wanted to know if anything similar could happen in closely monitored National Health Service hospitals. To explore this idea, he took the real records of two of his colleagues and randomly switched some of the outcomes from successful to unsuccessful, increasing the rate at which their patients appeared to die to make it about the same as Shipman's. That is, he generated dark data to show the sort of thing that would happen if someone like Shipman was at large in an NHS hospital. Nashef's study shows the power of the approach: "The results of the experiment were outstanding. . . . Dr. Harold Shipman went undetected for 25 years. . . . In our experiment John, the anesthetist, was picked up after ten months, and Steve, the surgeon, was picked up even quicker, within eight months."

Perhaps a more familiar application of simulation is in machines like flight simulators. These are used to train pilots, who can be subjected to extreme and unexpected situations in the simulation without the risk of crashing a real plane. Again, these artificial situations represent data which could have arisen.

To explore the ideas in more detail, let's look at simulation to explore a situation in which we can easily work out the mathematics, namely, the behavior of a tossed coin.

Using some elementary statistics I can calculate the probability that a fair coin tossed 10 times will come up heads fewer than five times: it's the probability that it comes up heads four times plus the probability that it comes up heads three times plus the probability . . . all the way down to the probability that no heads at all come up. And that total probability is 0.377. That's all very well, but doing that calculation requires knowledge of the binomial distribution of statistics. However, there is an alternative way to estimate that probability. We could toss a coin 10 times and actually see how many heads came up. Doing this once would not be sufficient—either fewer than five coins will come up heads or they won't, so it will give us a yes/no answer rather than an estimate of the probability that fewer than five heads will come up. We need to repeat the exercise of tossing 10 coins multiple times to see *what proportion* of those times fewer than five heads are produced. Indeed, to get an accurate estimate we need to repeat the exercise many times (and the more the better; the law of large numbers, mentioned in chapter 2, tells us this). But this process could rapidly become rather tedious. So, instead, we simulate the coin tossing on a computer by randomly generating 10 values, each 0 or 1 (0 corresponding to tails and 1 to heads), with probability 1/2, and see if we get fewer than five 1's. Then we do it again. And again, and again, and . . . and see what proportion of those times yields fewer than five heads.

I ran this simulation a million times. The proportion of those million which came up with fewer than five heads was 0.376, not far off the true value for the probability. It's worth drawing attention to the "million" there. Simulation has really come into its own with the power of the modern computer.

That was a straightforward example, one which I could run on my laptop and indeed one to which I knew the correct answer. At the other extreme we find weather and climate simulations, based on vast data sets and using the most powerful computers available. They use extremely sophisticated models of the interlocking processes affecting climate: models of the atmosphere, of ocean currents, of solar radiation, of biological systems, of volcanic action, of pollution, and other influences. The reactive nature of these systems means they pose particular challenges: push a ball and it rolls away from you, but push a complex system and it can react in unexpected and often unpredictable ways. The scientific use of the word *chaos*, as in chaos theory, owes much to the fundamental unpredictability of meteorological systems. Equations describing such complexity often cannot be solved to provide clear and crisp answers; they have an intrinsic uncertainty. Instead, simulation comes to the rescue, repeatedly generating data from the models to show how weather and the climate might behave. The results of these simulations will give us an idea of the range of behaviors that can result, such as how often extreme events, like floods, hurricanes, and droughts will appear. Each run of such a simulation yields data which could have arisen but data we did not actually see and which we might describe as dark.

The same sort of approach is used in economics and finance. As you might expect, models of a modern economy are pretty complicated. Society is made up of millions of people all interacting with others and pulling in their own directions, organized in a wide variety of social structures, and impinged upon by external forces. Writing and solving mathematical equations to see how such systems should be expected to change over time is clearly challenging. Simulation, by generating data which could have arisen in such a society, allows us to explore how such masses of people are likely to evolve and how they will react to changes such as the imposition of trade tariffs, breakout of wars, adverse weather conditions, and so on.

Simulation is also used in more subtle ways in modern data analysis. In particular, modern so-called Bayesian statistics, described later in this chapter, often lead to very complex and intractable mathematical equations. Finding solutions to these equations is difficult or impossible, and alternative methods have been developed based on simulations. That is, as in the climate example, we take the equations as a model and generate data which could have arisen if that model were true. And then, as before, we repeat this time after time, to generate multiple data sets, each of which could have arisen. It is then relatively straightforward to summarize these multiple data sets—to calculate averages, ranges of variation, or any other descriptive statistics we want. These tell us what characteristics the outcomes are likely to have and how likely they are to arise. These sorts of simulation methods have transformed Bayesian statistics from an interesting theoretical idea to a practically useful tool—one which underlies much machine-learning and artificial intelligence work.

However, it is important to remember that, by definition, simulated data come from a hypothetical model of the underlying process. They are fabricated synthetic data (*DD-Type 14: Fabricated and Synthetic Data*) and don't arise from the real process itself. Obviously, then, if the model is wrong—if it is a poor representation of reality—there is a risk that the simulated data are poor representations of what might have happened. But once again that's generally true—if you do not have a good understanding of something, then you can be led astray.

Replicated Data

As we have seen throughout this book, we often find ourselves wanting to estimate the value of something which has not been or perhaps cannot be observed directly. For example, we might want to determine which of two diseases someone has, based solely on their symptoms; or to predict the number of passengers who will travel through New York's subway system next year, based on how many traveled this year plus what we think next year's national economy will be like; or to predict whether someone will default on a loan; or a student will succeed in his or her course; or a potential employee will do an effective job; and so on.

The common feature of these situations is that we have data describing a set of previous cases (people who have had the diseases, previous years' data on subway travelers, or the behavior of people who previously had a loan, for the preceding examples). And for each of these cases we know both the outcome (which disease, how many travelers, whether they defaulted), as well as the descriptive characteristics (the symptoms, journey patterns, application form details). We can use these past data to model the relationship between the descriptive characteristics and the outcome. That model will then enable us to predict the outcome for other cases, based solely on their pattern of descriptive characteristics.

This basic structure—a collection of past cases for which we know both their characteristics and their "outcome," which we wish to use to construct a model relating characteristics to outcome so that we can predict the outcome of new cases—is ubiquitous. The models are often termed *predictive* models—though the "prediction" might just be of the unknown diagnosis, as in the disease example, not necessarily of the future, as in the subway example. Because such situations are so common, they have been the focus of a tremendous amount of research. A large number of different approaches to building such models have been developed, with a wide range of different properties, so that some methods will be better for some problems and other methods for other problems.

But what has all this to do with dark data? We'll illustrate with a very simple example and the most basic kind of predictive approach. Our example will be to predict income from just a single variable, age. To construct a possible model we will collect data on age/income pairs of values from a sample of people. Then, a very basic approach would be to predict the income of a new person, whose age we knew, by using the income of others of the same age. So, for example, if we wanted to predict the income of someone aged 26, and there was one person aged 26 in our sample, then at the simplest we'd use their income alone as our prediction. If there were additional 26-year-olds in our sample, we'd want to use the information from all of them, so we would use their average income. More generally, averages will provide a better prediction, since they are less susceptible to random fluctuation. That means it might also be sensible to include the incomes of any 25-year-olds and 27-year-olds, since they are likely to be close to those of the 26-year-olds, and including them will increase the sample size. Likewise, we might want to include those aged 24 and 28, and so on, but giving less weight as we move further from 26. This strategy would also give a prediction even if there was no one aged exactly 26 in the sample of past data.

To see why this can be thought of as involving dark data, here's a different perspective on the same exercise. To predict the income of someone aged 26, we create a new set of data by randomly replicating values in the existing sample. We make many replicates of any aged 26, fewer replicates of those aged 25 and 27, fewer still of those aged 24 and 28, and so on. It's as if we originally had a much larger sample, most of which we hadn't seen before. Averaging the incomes of all this created data will give us an appropriate estimate of the income of people aged 26.

That example is all very well, but in real life things are usually more complicated. Typically, instead of just one descriptive characteristic (like age in the example), we will have several—or even many. For example, we might characterize patients by their age, height, weight, sex, systolic and diastolic blood pressure, resting pulse, and a number of symptoms and medical test results, and want to estimate the probability that a new patient with a particular set of characteristic values is likely to recover. Following the preceding example, we would create a new data set, replicating people in such a way that we have more replicates of people who have very similar characteristics to the person we were interested in, and fewer replicates the more different they are. People who are completely different from the patient of interest (they are male rather than female, young instead of old, and show a completely different symptom pattern from the person for whom we want to predict the illness probability) might not be replicated at all. Once we have produced all this replicated data, we simply look at the proportion among them who recover, and that proportion is our estimated probability of recovery.

That basic idea of strategically replicating data to produce larger, more relevant data sets has been used in other ways too. To keep things simple, we'll again look at machine-learning algorithms for assigning objects to classes, as in the preceding diagnostic example, or to decide whether a mortgage applicant is likely to default on repayments (two classes: is or is not likely to default), and so on. But now we'll examine how to use the ideas to improve the performance of such algorithms.

Typically, algorithms for making such diagnostic classifications will get some wrong: symptom patterns are often ambiguous, and a young applicant might have only a short history of financial transactions. One way to improve such an algorithm is obviously to look at the cases it previously misclassified and see if we can amend or adjust it in some way to improve its predictions so that it gets more of those cases right. And one way to do this is to create fictional data, as follows. We begin by identifying the patients or applicants (for example) for whom we made the wrong prediction and we *add extra copies of these cases*, perhaps a great many extra copies, to the data. Now, when we tweak the parameters or structure of our model to classify this expanded data set it will be forced to pay more attention to the cases it previously got wrong. To see this, imagine the extreme situation in which a previously misclassified case has been replicated 99 times, so that now there are 100 identical copies of it. Previously, our classification got just one version of it wrong when it misclassified this case. But now, since the 100 cases are identical, it will get 100 wrong. The performance of the classification method will be much improved if we can adjust it so that it now gets this case—*and its 99 copies*—right.

In short, modifying the algorithm by applying it to this new data set—the original data enhanced by a large number of copies of those previously misclassified—will lead to a new version which now more accurately classifies those data points that were previously misclassified. The idea is to create data which shift the algorithm's "attention" in the direction we want it to go. Or, as we might think of it, we use data which might have been.

This idea is called *boosting*. It was revolutionary when first proposed but is now widely used in machine-learning. Indeed, at the time of writing, variants of boosting are generally front-runners in machine-learning competitions run by organizations like Kaggle. (For example, a sophisticated variant called *extreme gradient boosting* is doing very well.)

Whereas boosting focuses on the cases which were previously misclassified, since these are clearly the cases where extra attention is needed, a different approach to using fictional data sets has been developed for determining the accuracy of estimates. This is the idea of *bootstrapping*, invented by the American statistician Brad Efron. (Statisticians and machine-learning experts are good at devising colorful names which reveal the essence of the concepts.)

Here is how bootstrapping works.

As we have seen, often our aim is to make an overall summary statement about a population of numbers (e.g., we might want to know the average value), but sometimes it is impracticable to determine each individual number. For example, we might want to know the average age of the people in a country, but perhaps there are too many people in the country for us to ask everyone their age. Indeed, as we have seen before, sometimes the very notion of taking "all possible measurements" makes no sense—we could not repeatedly measure the weight of a rock an infinite number of times, for example. The answer, as we have seen, is to take just a sample—to ask just some people their age, and to take just some measurements of the rock's weight—and then to use the sample average as our estimate.

Now, sample averages are useful—they certainly do give us an idea of the overall value—but it would be unrealistic to expect them to be completely accurate. After all, had we chosen a different sample of data—a different set of people to ask their age or a different set of measurements of the rock's weight—we would probably have obtained a different result. We would hope it would not be too different, but it would be unrealistic to expect it to be identical. This means that in addition to the average, we also want some measure of its accuracy. We want to know how much variability there is in the average values we would obtain if we drew different samples and how far from the true value our obtained sample average is likely to be.

For averages, finding such a measure of variability is straightforward using well-understood statistical theory. However, for other descriptions and summaries of the data, especially if we're doing things far more complicated than simply finding an average, it is much tougher. But synthetic dark data can again come to the rescue.

If we could draw many samples (like repeatedly tossing a coin 10 times, as in our earlier example), we wouldn't have a problem: we'd just draw the samples, fit a model (e.g., calculate an average or do a more complicated calculation) for each sample, and look to see how much variation there was between the results, that is how different they were. But unfortunately, we have only one sample.

Brad Efron's insight was to suggest thinking of the one sample we have as if it were the entire population. Then, in parallel with the idea of drawing a sample from the population, we could draw a *subsample* from the sample (each subsample being the same size as the original sample by permitting each value to be drawn multiple times). In fact, just as we could (in principle) draw lots of samples from the population, so we can draw many subsamples from the one sample we have. The difference is that we can actually draw these subsamples. For each of these subsamples we can then fit a model (e.g., estimate the average of the values in each subsample), and then we can see how much variation there is between these averages. The idea is basically to model the relationship between the population and the actual sample by the relationship between the sample and a subsample. It is as if we are creating fictional replicates of the sample—as if we are revealing a large amount of data which had hitherto been concealed. The word *bootstrap* comes from moving from sample to subsample in just the same way as we move from population to sample, so that it's as if we reached down and "picked ourselves up by our own bootstraps."

One thing is very evident from these examples of how simulated data can be used to aid inference and prediction: they are all very labor intensive. Creating copies of nearby data values, replicating misclassified cases, drawing multiple (often hundreds or more) subsamples of data, are not things to be undertaken lightly. Or, at least, they wouldn't be if we had to do them by hand. Fortunately, however, we now live in the world of computers. These are designed to effortlessly undertake repetitive calculations in fractions of a second. Recall the *million* tosses of 10 coins I ran for the coin-tossing example. The methods described here, creating dark data to use to our advantage, are very much children of the computer age. I used the word "revolutionary" earlier. That is no exaggeration.

Imaginary Data: The Bayesian Prior

When we looked at simulation we assumed we had a proper understanding of the underlying structure, process, or mechanism which produced the data. Such a degree of confidence is often unjustified. We might have an *idea* about the process, but we would be lucky if we knew everything exactly. In particular, we might be unsure about the precise values of the numbers characterizing the structure.

For example, I might believe that the heights of men in the British population follow what's called a *normal distribution,* with most being around the average height and few very tall or very short, but I might not know what that average is. I might be pretty sure that it's less than 6 feet, more confident that it's less than 6 feet 1 inch, even more confident still that it's less than 6 feet 2 inches, and so on, with no doubt whatsoever that it's less than 7 feet. Likewise, I might be fairly confident that it's more than 5 feet 6 inches, more confident that it's greater than 5 feet 4 inches, and very sure that it's more than 5 feet 2 inches. What I'm describing here is a distribution of my belief in the likely value of the average height; I'm saying how confident I am the average height lies in any given range.

That belief distribution has come from somewhere, although it might be difficult to say precisely where. It might be a combination of awareness of the heights of the people I've met in the past, vague recall of previous studies I've read, a distant recollection of what someone told me in the past, or any other source. In any case, it is equivalent to an earlier data set, even though I will almost certainly be unable to pin those data down and give values for them; they are dark data.

Given the fundamental subjectivity and uncertainty about where our beliefs about the average height of the population come from, we might be hesitant about making statements or decisions based on what we *think* the average might be. Instead, we would like to collect data to gain some objectivity. And this is exactly what the Bayesian approach to statistics does. It begins with initial beliefs about the possible values of the average height, called our *prior* beliefs, and then adjusts those beliefs as new *real* data become available, resulting in our *posterior* beliefs. We might, for example, measure the heights of a random sample of 100 British people and then use those 100 numbers to adjust or update our initial belief about the overall population average. The result will be a new distribution of possible values for the average height, in which our earlier belief distribution is shifted toward the values we actually observed. If we took a very large sample, the shift would be almost completely toward that sample, leaving very little influence from our initial subjective beliefs. This updating or adjustment process is made using a fundamental theorem of probability called *Bayes's theorem*. In our terms, Bayes's theorem combines unobserved dark data with real observed data to yield a new distribution of beliefs about the likely average height. (For what it's worth, the UK's Office for National Statistics gives the average height of UK males as 5 feet 9 inches.)

Here's an example. Researchers began trying to determine the speed of light as early as the seventeenth century: in 1638

Galileo determined that it was at least 10 times the speed of sound, in 1728 James Bradley estimated it to be 301,000 kilometers per second, and in 1862 Léon Foucault estimated that it was 299,796 kilometers per second. We can summarize these and other estimates to yield a distribution of how confident we are in different possible values. The detailed experimental results may be lost—may be dark data—but the belief distribution captures the relevant information they contain. In the late nineteenth century, the Canadian American astronomer and mathematician Simon Newcomb (the same man we met earlier in connection with the Benford distribution) carried out further experiments. He recorded the measurements he made between 24 July 1882 and 5 September 1882 in a paper published in 1891 in the *Astronomical Papers* of the U.S. Nautical Almanac Office.[1] Newcomb's detailed measurements can then be combined with the dark data implicit in our belief distribution from the earlier experiments to yield an improved distribution. Our current single best estimate of the speed of light is that it is 299,792.458 kilometers per second in a vacuum.

Bayesian statistics has become hugely important—it is one of the two (or some would say, three) major schools of statistical inference.

Privacy and Confidentiality Preservation

Until this point in this chapter we have examined existing statistical procedures and philosophies from the dark data perspective, rather than from the perspective of the observed data. This way of looking at things can often lead to new insights. But there are also other ways in which we can take advantage of dark data. In fact, as we shall see, concealing data is central to the effective functioning of modern society: many of our ordinary everyday activities would be impossible without hiding data.

In chapter 6, and to some extent in chapter 7, we saw how fraudsters hide information. Their aim is to present a distorted impression of what is going on: to make you believe you will benefit from a transaction, whereas in fact you will lose, or to conceal the truth about the results of an experiment. Spies do the same sort of thing. Their real objective, probably their real identity, certainly their real behavior, is concealed from their target (a government or corporation, perhaps). Spies don't want opposing governments to know what they are up to, so they endeavor to keep their activities hidden. And contrariwise, spies are trying to discover and pass on data which others don't want them to see: data which governments in turn are trying to keep secret from the spies. And as if all that wasn't convoluted enough, it may be advantageous at a higher level if the data the government is apparently trying to keep hidden are leaked, so that any enemies know their capabilities and do not become paranoid. And then there are double agents, in which case it may not be clear what is concealed from whom. Things can quickly get very complicated!

But fraudsters and spies are not the only people who want to keep things hidden. You may not be keen on your medical or financial history becoming public knowledge. You might be embarrassed if some aspects of how you live your life are generally known. This is the essence of privacy, and Article 12 of the *Universal Declaration of Human Rights* begins: "No one shall be subjected to arbitrary interference with his [sic] privacy. . . ."

Privacy itself has been variously defined. It is the right to be left alone and not be bothered; it is the right to be protected from government intrusion; and it is the power to selectively reveal ourselves to the world. These high-level definitions are all very well, but privacy and concealment also matter at a more practical level. You use secret passwords to protect your bank account, your social media accounts, your phone, your laptop, and so on.

This means the data in your accounts are kept dark from those you want to protect them from. It also means it is important to know how to create a good password.

A disturbingly large number of people still stick to default passwords, like "password," "123456," "admin," and so on. (Which brings to mind the old joke about the man who changed all his passwords to "incorrect," so that when he forgot it the system would say "your password is incorrect.") These are often the initial passwords associated with equipment when it's bought and which should be changed as the equipment is installed. Hackers typically start by looking at those default passwords. A more general basic hacking strategy used to break passwords is simply to try billions of different character combinations—which, with computers, can be done at a rate of half a million a second (again recall my generation of a million sets of 10 coin tosses in a matter of seconds). If the hacker has some information about your password—that it consists only of numbers, perhaps—then this dramatically accelerates the search. This is why you are advised to include upper- and lowercase letters, as well as numbers and special symbols. Doing so leads to a much larger dictionary of characters and a *much* larger space of possibilities which the hacker has to search through. With just 10 digits and a password eight characters long, there are 10^8, that is a hundred million possible passwords. Testing them at a rate of half a million a second means you can run through all of them in 200 seconds—a little over three minutes. On the other hand, using the 10 digits, and all 26 letters of the alphabet, both upper- and lowercase, along with say 12 special characters, means the number of possible passwords eight characters long is 74^8, which is about 9×10^{14}. At half a million a second it would take about 28.5 million years to go through them all. Phew!

Spying and passwords are closely linked to secret codes and cryptography. These are tools for communicating information between two people while preventing others from reading it—keeping them in the dark. In fact these ideas apply much more widely than just to spying. For example, commercial organizations often wish to exchange confidential information, banks need to know that no one has intercepted a message and changed or redirected it, and you might want to be confident that your email and other communications can be read only by the intended recipient and not by anyone able to hack into the system.

Cryptography, the science of coding, has become an advanced mathematical discipline. Modern methods are often based on *public-key cryptography*. This is an elegant mathematical solution which uses two numerical "keys," one of which allows people to encrypt messages and the other to decrypt them. The first is made public, so anyone can encode a message. The second is secret, so only those in the know, that is only those who have the private key, can decode messages.

The public, social, and ethical issues associated with cryptography can be deep. While many legitimate transactions depend on encryption for their security, so also can illegitimate transactions, such as criminal or terrorist communications. Apple Inc. has received a number of requests from U.S. courts to reveal information on locked iPhones. In one case the FBI asked Apple to unlock a phone owned by one of the perpetrators of the December 2015 San Bernardino, California, terrorist attack in which 14 people were killed. But such a request opens up wide questions on the extent to which personal privacy should be sacrosanct. In that particular case Apple declined, and a hearing was scheduled, but before it could take place the FBI found a third party who could unlock the phone, and the request was withdrawn. The story of privacy, secrecy, and access to dark data on mobile phones is a long and unfinished one.

National Statistical Institutes (NSIs) are official statistical bodies which collect and collate data from entire populations and then analyze them to yield an understanding of the population in order to develop effective social and public policies. Such bodies need to keep the personal details—the microdata—secret while allowing statistical summaries to be published. For example, you would expect your country's NSI not to divulge your salary or past medical history while being able to give information on the salary distribution and the numbers of people who had had different illnesses. This policy can lead to delicate privacy issues. In particular, if information about a small group of people is given out, it might be possible to identify individuals within that group. For example, publishing information on men aged between 50 and 55 and who live in a certain ZIP Code might narrow things down substantially. In the worst case there might be only one person who meets all the conditions defining the group.

Because of these delicate issues, NSIs and other bodies have developed tools for keeping data dark, so that information about populations can be given out without compromising any individual's privacy. For example, if a cross-classification of criteria leads to a small number of people (e.g., those who live in a certain town and earn over £1 million a year), then this cell of the cross-classification can be combined with neighboring cells (e.g., those who live in a group of local towns and earn over £100,000 a year).

Another strategy that NSIs use to conceal details of data is to randomly distort or perturb them. For example, a small random number might be added to each count of a table so that it can be published without divulging the real, exact numbers but in such a way that the big picture is still well represented. Indeed, there are ways of doing this so that aspects of the big picture (e.g., overall population averages, distributions of numbers in different groups) are preserved *exactly* while perturbing all the constituent numbers.

A third strategy is to model the distributions and characteristics of the true data and then use the model to generate synthetic data with the same properties—just as in the simulation approach discussed earlier. For example, we might calculate the average age and the spread of ages in a population, as well as the general shape of the age distribution, and then generate artificial data which has the same average, spread, and distribution shape. Thus, the real data completely vanish, but (up to a point) summaries can be generated from the synthetic data which match those of the real data.

Data can also be *anonymized*; that is, information which can identify individuals is destroyed. For example, names, addresses, and Social Security numbers will be wiped from the records. The downside of anonymization is that any *potential* for reidentifying to whom each record belongs is lost. Thus, in clinical trial records, anonymity might be preserved by stripping away the names and addresses, and the like, but if at a later date it is found that some of the patients are at serious risk, it would be important to be able to go back to identify them. In any case, for many businesses, keeping records which enable individuals to be identified is critical to their operation.

In cases like these we can use so-called *pseudonymization*. Instead of simply destroying the identifying information, pseudonymization replaces it with a code. For example, names might be replaced by a randomly chosen integer (e.g., replacing "David Hand" with "665347"). If a file is kept somewhere which matches identifiers to codes, then, should the need arise, the identity of individuals can be found.

In fact, in formal definitions of anonymization used by statistical offices there is typically a clause saying something like the anonymization process should protect "individuals from being reidentified by any reasonable means," because perfect anonymization can seldom be guaranteed, since data sets may be *linked* to others, as discussed in chapter 3. As we saw there, the linking of data sets has huge potential for improving the human condition. For example, linking food purchasing patterns with health data will provide valuable epidemiological insights. Linking education data from schools and colleges to employment and income data from tax offices will provide immensely helpful information for developing public policy. These sorts of linkage exercises are not hypothetical but are being increasingly conducted by a wide range of bodies throughout the world. But they can be successful only if those included in the databases can be confident that privacy and confidentiality will be preserved. The UK's Administrative Data Research Network (ADRN), which we looked at in chapter 4, overcame such privacy and confidentiality risks using a "trusted third party" method for linking data, which meant that no particular data holder ever held both identifiers and the linked data.[2] For two data sets, the system worked as follows:

1. Each database manager created unique IDs for each record in their data set.
2. These IDs and the associated identifying information (e.g., names), but no other data, were sent to a "trusted third party" over a secure link, which matched the IDs using the identifying information.
3. A linkage ID was then created for each record.
4. Files containing the linkage ID and the unique IDs for each record were sent back to the database holders.
5. The database holders then added the linkage ID to the records in their data set.
6. Finally, each database holder stripped away the identifying information (e.g., the names) and sent the records

> and their linkage ID to the researcher, who could link the data sets using the linkage ID without being aware of the identities of any of the records.

This process sounds complicated, but it's very effective in keeping the identities concealed when linking the records from the two data sets and absent from the linked set.

Although such linkage exercises can be extremely valuable for society, linked data always carries identification risks, and these can be dramatically magnified if the data might be linked with future *external* data sets. (That was not possible with the ADRN, because all analysis also occurred in secure environments without access to other data sources.) This dilemma is illustrated by a famous early case.

In 1997 the Massachusetts Group Insurance Commission (GIC) released hospital data for researchers to study so they could develop improved healthcare strategies. The then governor of Massachusetts, William Weld, reassured the public that their privacy was protected by pointing out that the GIC had deleted personal identifiers from the data.

However, the possibility of data linkage was not addressed. Latanya Sweeney, who was then a graduate student at MIT, was studying "computational disclosure control," the branch of computer science concerned with tools and methods for keeping data dark. Sweeney wondered if, contrary to Weld's claim, it was possible to identify individuals within the GIC data. In particular, she decided to search for Weld's data. It was public knowledge that Weld lived in Cambridge, Massachusetts, which had only seven ZIP Codes and 54,000 residents. Sweeney matched this information with data in the electoral roll for Cambridge, which was available for purchase for only $20. Then, using other publicly available information about Weld—his birth date and his sex for example—she was able to match his electoral roll record with the hospital record and thus identify his medical records. To drive home the point, she sent him a copy of the records.

In fact, there is something unusual about this case. Any data set might include individuals who are anomalous in some sense and so permit easy identification, but that does not mean the majority of people could be identified. In the present example, Weld was a public figure about whom much information was generally known. Moreover, his reidentification hinged on the accuracy of the electoral roll. Daniel Barth-Jones has explored this case in detail, showing its unusual nature.[3] Nevertheless, the illustration is disturbing—so much so that recent changes in legislation have made such reidentification more difficult.

Another early famous case involved web search data openly released by the web portal and online service provider AOL in 2006. To preserve anonymity, AOL stripped out the searchers' IP addresses and pseudonymized their user names, replacing them with the same randomly chosen identifier whenever they occurred. In this case it was not a postgraduate student who cracked the anonymity but a pair of journalists, who quickly matched identifier 4417749 to Thelma Arnold, a widow living in Lilburn, Georgia. The journalists did this by narrowing their search using information contained in her searches: that she searched for medical ailments for people with last name Arnold, that she searched for material related to dogs, and so on.

And then there's the infamous Netflix Prize case from 2006. This database contained comparative ratings of film preferences by half a million Netflix subscribers. A $1 million prize was offered for developing an algorithm which was 10 percent more accurate than the existing algorithm used for recommending films to subscribers. Again, all personally identifying information was stripped out, and identifiers were replaced by random codes. This time, two researchers from the University of Texas, Arvind Narayanan and Vitaly Shmatikov, cracked the anonymity. They wrote: "We demonstrate that an adversary who knows only a little bit about an individual subscriber can easily identify this subscriber's record in the data set. Using the Internet Movie Database (IMDb) as the source of background knowledge, we successfully identified the Netflix records of known users, uncovering their apparent political preferences and other potentially sensitive information."[4]

The foregoing are all early cases in which apparently anonymized data was identified. Such cases have led to the introduction of legislation and passage of laws to make future data sets more secure and to penalize attempts to break anonymity. But the painful truth is that either data are completely dark, and hence useless, or there is a crack through which light might leak.

Collecting Data in the Dark

We've seen that data identifying individuals can be anonymized when they are linked to other data sets, but it's possible to go even further. It's possible to make data dark as they are collected and as they are used in calculations—so they are never actually seen in the first place—but still able to be used to make discoveries and extract value. The following are some ways these things can be done.

First, *randomized response* is an old strategy for collecting sensitive personal information—such as data relating to sexual or dishonest behavior. To take an example, let's suppose we want to know what proportion of the population has ever stolen anything. Asking people directly is likely to yield distorted responses, as presumably there will be a tendency to lie and deny it. Instead, we ask each person to flip a coin, which we do not see. They are instructed that if the coin comes up heads, they must truthfully answer yes or no to the question, Have you ever stolen anything? while if it comes up tails they must simply answer yes. Now, for any one person, a positive answer means we will not know if he or she has stolen something or if the coin simply came up tails. But we will learn something overall. Since the probability that the coin will come up heads is 1/2, we will know the overall number who answered no is just half of those who really did not steal anything. So doubling this number tells us how many really had not stolen anything. Subtracting this value from the total number tells us how many have stolen something.

David Hugh-Jones from the University of East Anglia in the UK used a variant of this idea to explore honesty in 15 different countries.[5] He asked people to toss a coin (without his seeing the outcome), with a $5 reward if it came up heads. If everyone told the truth, we would expect about half the people to report that heads had come up. If the proportion claiming heads was greater than a half, it suggested that people were lying—and this is what Hugh-Jones used as a measure of honesty.

The randomized response strategy is a way to hide data as they are being collected. There are also ways to hide data as calculations are carried out. *Secure multiparty computation* is a way of aggregating information from a group without any of them learning anyone else's data. Here's a very simple example. Suppose I want to know the average salary of a group of neighbors, but they are all very sensitive about disclosing how much they earn. In that case, I ask each of them to split his or her salary into two numbers *a* and *b* such that the sum of *a* and *b* equals the salary. So someone who earns £20,000 might split it into £19,000 and £1,000, or into £10,351 and £9,649, or into £2 and £19,998, or even into £30,000 and −£10,000. It does not matter how they split their salaries—positive and negative numbers can be used—provided the two parts add up to their total salary. Then all the *a* parts are sent to someone who adds them all up—giving a total *A*, say. And all the *b* parts are sent to someone else (it's important that it be a different person), who likewise adds them up—to give a total of *B*. The final step is simply to add *A* and *B* and divide by the number of people, to give the average. Note that throughout this process no one learns anyone else's salary. Even the people who do the adding have no idea of the missing (*a* or *b*) part they don't see.

Secure multiparty computation provides a way of summarizing data from a population without any member of that population, or indeed anyone else, learning the values for individuals within the population. But in fact it's possible to go even further. Homomorphic computation allows you to encrypt your data, to darken them, and have someone else analyze the encrypted data to yield an encrypted result without their knowing what the data or result mean. Then you—the only one who knows how to decrypt encrypted values—can decipher the result. In practical terms this method dates to about 2009, and a paper by Craig Gentry from the IBM Watson Research Center, but the idea dates to the 1970s.[6]

The following is a trivial and highly artificial example of the idea—real applications use much more sophisticated methods.

Suppose we want to calculate the average age of the people in a club, but we do not have a computer powerful enough to do this. We want to ask someone with a sufficiently powerful computer to do the calculations for us, but we do not want that person to see the ages themselves. To do this, we start by "encrypting" each of the ages by adding a different randomly chosen number to each of them. We also calculate the average of all the random numbers we added. We now send the encrypted numbers—the sums of the original numbers and their random partners—to the person who is going to do the calculation. She adds up the encrypted numbers and sends their average back to us. It is not difficult to see that if we then subtract the average of the random numbers from her average, we will obtain the average of the ages of the people in the club.

Obviously this is a very simple example, and normally we would be interested in doing something more demanding than taking an average.

We've seen that data can be collected without being seen by those collecting them, and we've seen that data can be analyzed without those doing the calculations understanding what they are analyzing. More generally, this chapter has turned the concept of dark data on its head. Normally, dark data are a source of problems: they hide from us things we wish to know, potentially leading to distorted analyses and misunderstandings. But this chapter has described ways in which hiding data can be extremely valuable and how it can lead to improved estimates, better decisions, and even protection from criminals.

Chapter 10

CLASSIFYING DARK DATA

A Route through the Maze

A Taxonomy of Dark Data

We have seen many examples of dark data in this book, looked at why the data were dark and the consequences of their being dark, and what we might try to do about the difficulties caused by their being dark. However, situations are often complicated because data might be dark for multiple reasons. Here's an example.

The UK government's Behavioural Insights Team has been called the "Nudge Unit" by the media. This unit searches for small strategically applied policy changes (nudges) that can have big impacts on behavior. A recent report from the team said "several reports and media articles have noted that official statistics show a large decline in calorie consumption in the UK over the last 40 years. At the same time, we have seen the population gain weight over this period. How has weight gone up, if we are eating less? . . . One response is that physical activity levels must have declined, leading us to expend much less energy."[1]

That seems possible, if a bit surprising. It suggests that while we have been eating less, we have been getting proportionately even less exercise, leading to a gain in weight. However, the report concluded that such a conclusion was not plausible, pointing out that "the reported level of calorie consumption is too low to sustain our current weight even if we were only doing the minimum possible level of physical activity." It also says "these estimates for calorie intake fall below the commonly cited recommended daily allowance figures of 2,500 kilocalories for men and 2,000 kilocalories for women (for people of healthy bodyweight)." The problem, the team suggested, was one of dark data.

Food purchasing levels are estimated from the Living Costs and Food Survey (LCFS). Calorie intake is estimated from the National Diet and Nutrition Survey and Health Survey (NDN-SHS) for England. The Nudge Unit's report suggests that these surveys are underestimating food purchasing and calorie consumption. Regarding the LCFS, the report pointed to "studies [which] have estimated that the percentage of economic activity not captured by the LCFS has increased from 2% in 1992 to around 16% in 2008." When the unit adjusted the LCFS results to allow for this factor, it found that food consumption had been increasing since the 1990s. As for the NDN-SHS figures, they were adjusted using so-called doubly labeled water studies, "the gold standard for measuring energy expenditure." This adjustment suggested that "as a population we are consuming 30% to 50% more calories than the levels reported in official statistics."

All the foregoing does seem like a classic case of dark data. Calorie intake has *not* fallen; it just looks that way because of missing or misleading data. The report proposed five reasons for such underreporting, involving various kinds of dark data:

- increasing levels of obesity (since obese people are more likely to underreport their intake—*DD-Type 11: Feedback and Gaming*);

- increased desire to lose weight (since this is associated with increased underreporting—*DD-Type 11: Feedback and Gaming*);
- increased snacking and eating outside the home (*DD-Type 2: Data We Don't Know Are Missing*);
- falling survey response rates (*DD-Type 1: Data We Know Are Missing, DD-Type 4: Self-Selection*);
- a growing discrepancy between reference data (used to calculate calories) and true portion sizes or food-energy density (measurement error hiding the true values: *DD-Type 10: Measurement Error and Uncertainty*).

The Nudge Unit report explicitly identified multiple possible reasons for the dark data, but in many situations the possibility of multiple causes will not be so obvious. Moreover, teasing reasons apart so that appropriate steps can be taken to overcome the risks the dark data pose is often far from easy.

The first step must always be to be aware there might be dark data. Indeed, your default assumption should be that the data are incomplete or inaccurate. *That is the most important message of this book: be suspicious about the data*—at least until it is proved they are adequate and accurate.

Additionally, you need to be able to recognize situations especially vulnerable to problems of dark data, particular signs that invisible dark data are distorting what have been collected, and more general situations in which danger lurks. This book has tried to facilitate that process in two ways.

The first is via the examples throughout the book, which have provided illustrations of the ways dark data can arise. They show specific situations you should look out for. Of course, there will be countless other situations in other contexts not illustrated in this book, but it is hoped the examples here will serve as a starting point.

The second way the book has sought to help you recognize particularly risky situations is via the taxonomy of *DD-Types* of dark data, introduced in chapter 1 and referred to throughout. For ease of identification—and of use in real, practical situations—they are summarized next, with examples of each.

These *DD-Types* span the space of "kinds" of dark data, just as horizontal and vertical axes span the two-dimensional plane of a graph. However, in contrast to the axes of graphs, my *DD-Types* do not completely characterize the space of dark data. First, there will doubtless be causes of missing or inadequate data not mentioned here. And second, other new types of data are arising continually and these will have their own new types of darkness—as is discussed in the final section. Nonetheless, the aim of the list of *DD-Types* is to provide a partial checklist of dangers and the sort of general issues you should look out for in any data set and when undertaking any data analysis. But you should always bear in mind that merely because you manage to identify the presence of one *DD-Type* does not mean that others are not also present.

DD-Type 1: Data We Know Are Missing

These are Rumsfeld's "known unknowns." They arise when we know there are gaps in the data, concealing values which could have been recorded. An example is table values that are missing, as in the marketing data extract in Table 1, or failure of people on a list to respond to an interview, either in part or at all. In the latter case, perhaps all we know about the respondents who refused to take part is their identifying information.

DD-Type 2: Data We Don't Know Are Missing

These are Rumsfeld's "unknown unknowns." We do not even know that we are missing data. An example arises in web surveys, for which we do not have a list of possible respondents, so we do not know who has failed to respond at all. The *Challenger* space shuttle disaster represented an oversight of this kind, as the teleconference attendees did not recognize they were missing some data.

DD-Type 3: Choosing Just Some Cases

Poor choice of criteria for inclusion in a sample, or poor application of reasonable criteria, can lead to sample distortions. A researcher might choose healthier patients; an investigator might choose people sympathetically inclined to a company being evaluated. A particular variant arises when "the best" are chosen from a large number of cases, since this is likely to lead to disappointment in the future as regression to the mean kicks in. Likewise, p-hacking and failure to allow for multiple hypotheses means scientific results might not be reproducible.

DD-Type 4: Self-Selection

Self-selection is a variant of *DD-Type 3: Choosing Just Some Cases*. It arises when people themselves can decide whether to be included in a database. Examples are nonresponse in surveys when the respondents choose whether or not to answer questions, patient databases for which patients can decide whether to have their data stored (opt in and opt out), and more generally in the choice of services that people make (e.g., a bank or supermarket). In all these examples, those included may well differ in some systematic way from those not included.

DD-Type 5: Missing What Matters

Sometimes a critical aspect of a system is entirely unobserved. This can lead to mistaken causal attributions, as when an increase in ice-cream sales is followed by grass drying out. Obviously, here the causal network is missing data about the weather—but it's not always so obvious that something is missing. A more troublesome example is Simpson's paradox, in which an overall rate can increase while *all* constituent rates decrease.

DD-Type 6: Data Which Might Have Been

Counterfactual data are the data we would have seen had we taken some other action or observed what happens under different conditions or circumstances. An example is a clinical trial in which each patient can receive only one treatment—perhaps because the aim of the trial is to investigate time to cure—so that once a patient has been cured, it is not possible to go back to explore the time the alternative treatment would have taken. Another example is the age of the spouse of someone who is unmarried.

DD-Type 7: Changes with Time

Time can hide data in many ways. For example, data might no longer be an accurate description of the current state of the world, cases might not be observed because they occur after the end of the observation period, cases might drop out because they change their nature, and so on. Examples include medical studies of survival times after diagnosis and when the observation period is terminated before a patient has died, and data describing a country's population 20 years ago, which might be of limited value for developing current public policy.

DD-Type 8: Definitions of Data

Definitions might be inconsistent and may change over time to better reflect their purpose and use. This can cause problems with economic (and other kinds of) time series, in which the data underlying them may cease to be collected. More generally, if people define concepts in different ways, they may well draw different conclusions.

One example is UK crime statistics, which are measured by police records and by a survey of victims, since the two sources have different definitions of a crime.

DD-Type 9: Summaries of Data

By definition, summarizing data means discarding the details. If you report merely an average, you reveal nothing about the range of the data, and nothing about the skewness of the distribution. An average could conceal the fact that some values are very different indeed. Or, at the other extreme, it could conceal the fact that all the values are identical.

DD-Type 10: Measurement Error and Uncertainty

Measurement error leads to uncertainty about the underlying true value. This is most easily seen if we imagine a situation in which the range of measurement error is as large as or larger than the range of underlying true values, since then the observed value can be very different from the truth. Rounding, heaping, ceiling, floor effects, and others all inject uncertainty into the data, obscuring precise values. A different cause of uncertainty and inaccuracy is data linkage, in which identifying information might be stored in different styles, making matching error-prone.

DD-Type 11: Feedback and Gaming

This type of data arises when the values of data which have been collected influence the collection process itself—as in grade inflation and share-price bubbles. It means that the data are a distorted representation of the underlying reality—possibly drifting further from it as time progresses.

DD-Type 12: Information Asymmetry

Different data sets may be held by different people, and information asymmetry arises when one knows something that others don't. Examples are insider trading, Akerlof's market for lemons, and international tensions arising from limited knowledge of an enemy country's capabilities.

DD-Type 13: Intentionally Darkened Data

This particular example of choosing just some cases is a particularly troublesome one. It arises when people deliberately conceal or manipulate data with intent to deceive or mislead—it is fraud. We saw that it can arise in many contexts and in many ways.

DD-Type 14: Fabricated and Synthetic Data

When data are made up it might be with the intention to mislead, as in fraud. But it also occurs in simulation, when artificial data sets which *could have arisen* from the process being studied are generated, and in other applications in which data are replicated, such as bootstrap, boosting, and smoothing. Modern statistical tools make extensive use of such ideas, but poor replication can result in misleading conclusions.

DD-Type 15: Extrapolating beyond Your Data

Data sets are necessarily always finite. That means they have a maximum and a minimum value, beyond which lies the unknown. Making statements about possible values above the maximum or below the minimum in a data set requires that assumptions must be made, or that information is acquired from some other source. We saw an example of this with the *Challenger* disaster, where the launch occurred at an ambient temperature below any previously experienced.

Illumination

It would not be stretching the truth to suggest that in recent centuries the march of civilization has been concurrent with the advance of data science. The word *data* is, after all, almost synonymous with the word *evidence,* and this is what has been at the heart of the technological progress and enlightenment which has grown economies and advanced societies over the past few hundred years.

Indeed, paralleling the way modern industrial development has been powered by fossil fuels, data have been described as the "new oil." As with oil, vast fortunes have been made by those who can control and manipulate data effectively. More important, also like oil, data have to be refined—cleansed and preprocessed—to make them useful. Coping with the contamination of dark data is an example of such cleansing.

But the analogy is a rather strained one. Unlike the worth of oil, which can be used by anyone, the value of data depends on what you want to know. Moreover, also unlike oil, data can be sold or given away without your relinquishing it. Indeed, data can be copied and reproduced an unlimited number of times. And, of course, data can be dark: the data you don't have might make the data you do have of very limited value. Beyond that there are issues such as privacy and confidentiality which simply have no analogy in the world of oil. Data are far more than merely another commodity, which is precisely why governments are struggling so hard to sort out issues of data governance and ethics.

In very large part the data revolution has been driven by *observational* data. As we saw in chapter 2, these are data describing the natural evolution of some process, without any explicit intervention, and, as we have also seen, they are particularly vulnerable to the risk of dark data. Observational data are contrasted with experimental data, in which the levels of various factors are controlled. Additionally, vast quantities of observational data, often of new kinds, are being generated by automatic data collection systems and as a side effect of various administrative operations.

A now-classic example of the insights to be derived from a new kind of data is given by the *Billion Prices Project*. Alberto Cavallo and Roberto Rigobon from the Sloan School of Management scraped a huge number of online prices from the web and used these to construct inflation indexes. Using this source of data they showed how the level and dynamic behavior of inflation trends in Brazil, Chile, Colombia, and Venezuela could be approximated. But then they went further to note that "in Argentina, by contrast, there is a large unexplained difference in the level of online and official inflation rates."[2] There appeared to be no simple explanation involving which data were collected or how they were analyzed that could explain this discrepancy. Cavallo concluded: "The results for Argentina, by contrast, confirm the suspicion that the government is manipulating the official inflation series. It is the only country where online inflation deviates significantly from official estimates over time."

Cavallo and his colleagues had entirely sidestepped the traditional data-collection exercise for inflation indexes. As we saw in chapter 3, this involves employing teams of researchers to go out into shops noting the prices of goods on display. Not only is this expensive, but it is also slow. In contrast, the Billion Prices Project can produce an index which can be updated daily.

There is no doubt this is a tremendous big-data success story. But even so, things might not be as straightforward as they initially appear. Cavallo and Rigobon commented: "We . . . focus almost exclusively on large multichannel retailers and tend to ignore online-only retailers (such as Amazon.com)."[3] They drew attention to the fact that online prices cover a much smaller set of retailers and product categories than covered by the traditional approach to price indexes and noted it was necessary to decide which websites to collect data from, with the obvious danger of the smaller sites vanishing as dark data. Moreover, online prices are merely prices and do not indicate how much of each product is sold.

The point is not that these are insuperable obstacles. Indeed, having noted them, we can work toward overcoming them. Rather, the point is they imply dark data, so that the notion of inflation constructed by the Billion Prices Project is subtly different from the conventional definitions.

There are other, possibly more serious challenges of dark data in web-scraped data sets. For example, Google's search algorithm is constantly being updated to make it more effective, but the details of these changes are typically opaque to everyone except those deeply involved in making them. Recent changes include the introduction of a quality score of web pages when ranking, down-ranking of sites deemed manipulative, natural-language processing to better match the *intent* of the query, up-ranking mobile-friendly pages for mobile devices, and the identification of websites that violate Google's guidelines. All these changes seem sensible and beneficial, but the point is they change the nature of the data collected; that is, it is difficult to make comparisons between data collected before and after the change (*DD-Type 7: Changes with Time*). In particular, indicators of economic and social well-being might change in value not because the underlying reality has changed but because the data being collected about that reality have changed—so-called indicator drift. And dark data lie at the root of these changes.

We have seen that an entire domain of success stories arises from linking data sets, through combining, fusing, or merging data from different sources. The potential power of such exercises is obvious, since it is likely that different sources of data will contain information on different aspects of the objects being studied. Most commonly, these are people, and clearly such projects can be immensely useful for understanding and enhancing the health and well-being of society. But the risks of dark data from data linkage exercises are always very real. The populations in databases often do not match exactly (one might include some cases that the other does not), mismatches are frequently made during the matching process because of differences in the way data are stored (is John Smith the same person as John W. Smith or J. W. Smith?), and there might be duplicate records.

This book is primarily about how dark data can deceive humans (and then what to do about it). But they can also deceive machines. As applications of machine-learning and artificial intelligence become increasingly widespread, so we should expect to read more accounts of how dark data have misled the machines—causing mistakes and even accidents. In fact, there's a concept in machine-learning and computer vision sometimes called a "horse," named after Clever Hans.

Clever Hans was a horse owned by schoolteacher Wilhelm von Osten which could apparently do arithmetic. Hans appeared to be able to add, subtract, multiply, and divide, as well as to carry out more complicated operations such as telling the time, and seemed even to be able to read and understand German. Von Osten would ask Hans a question—written or spoken—and since Hans could not speak or write (it was not *that* clever), it would answer by tapping its hoof the right number of times.

Oskar Pfungst, a biologist and psychologist, investigated Hans in 1907. He came to the conclusion that while there was no trickery, Hans did not actually do the calculations. What Hans did was pick up on subconscious cues from his human trainer—who was able to solve the problems. Interestingly, the trainer did not know he was providing these cues. There's a similarity to tells in poker players.

The point is that Hans was not responding to what spectators thought he was responding to. And the same can happen with machines: their analyses, their classifications, their decisions might be based on unsuspected aspects of the input data, even on aspects of which you are unaware. In some cases, "imperceptibly small perturbations to a correctly classified input image [can mean] it is no longer correctly classified."[4] Using this weakness of automatic algorithms, researchers from Carnegie Mellon University have developed patterned eyeglass frames which look normal to us but which lead machines to mistake who their wearers are.[5] Disturbingly, researchers have found that the confusion is not limited to a particular neural network algorithm but is a general problem for that class of algorithms. Clearly, the machines are seeing things we are not, and equally clearly, those things do not represent an aspect we are interested in.

As we have seen over the course of this book, dark data can arise in an unlimited number of ways. They can be accidental, but they can also be deliberate: sometimes people attempt to hide the truth by presenting material in a certain way. Vigilance might be able to detect this, and a generally useful strategy is to look at the data from a different angle. A food might be described as "90 percent fat free," which could sound marvelous—but maybe the equivalent alternative, "10 percent fat," does not sound so good. In a similar vein, a treatment or lifestyle choice might be described as halving your risk of contracting some illness. But if the numbers showed that this halving was a reduction of the risk from 2 percent to 1 percent, you might not be that interested. Both might be regarded as so small as to be inconsequential. This concept is perhaps made even clearer by turning the figures around, so that the light shines more brightly on them: increasing the chance of avoiding the illness from 98 percent to 99 percent is perhaps unlikely to stimulate much interest.

The future is an obvious source of dark data. Despite the claims of various seers, clairvoyants, and prophets, the future is an unknown land, and unexpected events can always intervene to knock you off the path you thought you were following. Countless business failures provide very concrete evidence of this. The hedge fund Long Term Capital Management (LTCM) was already stressed when it was hit by the surprise Russian default on domestic local currency bonds in 1998. LTCM's potential collapse, with implications for a possible chain reaction leading to vast losses throughout the financial markets, led to the organization of a monumental bailout. Likewise Swissair, previously regarded as a very sound company, began an "aggressive borrowing and acquisition strategy" in the late 1990s—until the bottom fell out with the 2001 New York terrorist attacks, leaving it unable to manage the debt.

The following is a final example of a very literal kind of dark data, leading to yet another business collapse.

In the late 1970s and the 1980s a so-called format war took place between two formats of video recording: Sony's Betamax and JVC's VHS. In principle, Betamax was the superior technology, with higher resolution and better images, but VHS won. Betamax's technical superiority could not outweigh a greater cost and, initially at least, a recording time limited to one hour. In contrast, the initial VHS machines ran for two hours. The crunch was that Hollywood movies typically ran for more than one hour, which meant that a one-hour recording time would omit crucial data from the end of movies! Sony responded to this weakness by developing its technology to extend recording time, but by the time this was available it was too late: VHS had already gained the greater market share.

We are entering a brave new world. The possibilities for improving the human condition by enhancing our understanding and making reliable predictions based on data are limited solely by our imaginations. But we must tread carefully. Every step risks unexpected, concealed pitfalls. As I noted at the start of this book, we do not and *cannot* know everything about the data we are analyzing, about how they were collected or where they came from. And worse still, we don't know what we don't know. Since what we don't know might be critically important, one misstep can cause our understanding to fail and our predictions to be wrong—with possibly dramatic consequences for health, wealth, and well-being in general. The enthusiasm for data science is entirely justified, but it has to be tempered with caution. The only solution is to understand the risks and to exercise constant vigilance.

You are probably familiar with the old joke about the drunk looking for his keys beneath a lamppost, not because he dropped them there but because that's the only place with enough light to see. That encapsulates the risk of dark data clearly. Researchers, analysts, and indeed anyone trying to extract meaning from data are like that drunk if they restrict themselves to the data they have. Unless they understand how the data arose and what might be missing, they are at serious risk of looking only where they can see, not where the answers might lie. But we have traveled beyond the simple notion that dark data are merely data which could have been but were not recorded. We have also gone far beyond the distinction between known unknowns and unknown unknowns. Dark data can be all these things, but they can also be data which could not possibly have existed, or even data which we made up. The dark data perspective inverts the normal way of looking at things, leading to simplifications and deeper understanding when the data which are observed are embedded within a broader context which includes dark data.

It is hoped the countless situations involving dark data explored in this book will have raised your awareness of the risks, of what to be on the lookout for, and of methods of identifying dark data and correcting for them—ways of extending the patch of light out from beneath the lamppost to the surrounding areas. And hopefully they will also have enabled you to recognize situations in which strategic casting of shadows can lead to greater understanding and greater illumination.

NOTES

Chapter 1: Dark Data

1. https://blog.uvahealth.com/2019/01/30/measles-outbreaks/, accessed 16 April 2019.

2. http://outbreaknewstoday.com/measles-outbreak-ukraine-21000-cases-2019/, accessed 16 April 2019.

3. https://www.theglobeandmail.com/canada/article-canada-could-see-large-amount-of-measles-outbreaks-health-experts/, accessed 16 April 2019.

4. E. M. Mirkes , T. J. Coats, J. Levesley, and A. N. Gorban, "Handling missing data in large healthcare dataset: A case study of unknown trauma outcomes." *Computers in Biology and Medicine* **75** (2016): 203–16.

5. https://www.livescience.com/24380-hurricane-sandy-status-data.html.

6. D. Rumsfeld, Department of Defense News Briefing, 12 February 2002.

7. http://archive.defense.gov/Transcripts/Transcript.aspx?TranscriptID=2636, accessed 31 July 2018.

8. https://er.jsc.nasa.gov/seh/explode.html.

9. https://xkcd.com/552/; The Rogers Commission report on the *Challenger* disaster is available at https://forum.nasaspaceflight.com/index.php?topic=8535.0.

10. R. Pattinson, *Arctic Ale: History by the Glass*, issue 66 July 2012, https://www.beeradvocate.com/articles/6920/arctic-ale/, accessed 31 July 2018.

Chapter 2: Discovering Dark Data

1. D. J. Hand, F. Daly, A. D. Lunn, K. J. McConway, and E. Ostrowski, *A Handbook of Small Data Sets* (London: Chapman and Hall, 1994).

2. D. J. Hand, "Statistical challenges of administrative and transaction data (with discussion)," *Journal of the Royal Statistical Society, Series A* **181** (2018): 555–605.

3. https://www.quora.com/How-many-credit-and-debit-card-transactions-are-there-every-year, accessed 24 August 2018.

4. M. E. Kho, M. Duffett., D. J. Willison, D. J. Cook, and M. C. Brouwers, "Written informed consent and selection bias in observational studies using medical records: Systematic review,"*BMJ* (Clinical Research Ed.) **338** (2009): b866.

5. S. Dilley and G. Greenwood, "Abandoned 999 calls to police more than double," 19 September 2017, http://www.bbc.co.uk/news/uk-41173745, accessed 10 December 2017.

6. M. Johnston, The Online Photographer, 17 February 2017, http://theonlinephotographer.typepad.com/the_online_photographer/2017/02/i-find-this-a-particularly-poignant-picture-its-preserved-in-the-george-grantham-bain-collection-at-the-library-of-congres.html, accessed 28 December 2017.

7. A. L. Barrett and B. R. Brodeski, "Survivorship bias and improper measurement: How the mutual fund industry inflates actively managed fund performance" (Rockford, IL: Savant Capital Management, Inc., March 2006), http://www.google.co.uk/url?sa=t&rct=j&q=&esrc=s&source=web&cd=1&ved=0ahUKEwiavpGPz6zYAhWFJMAKHaKaBNQQFggpMAA&url=http%3A%2F%2Fwww.etf.com%2Fdocs%2Fsbiasstudy.pdf&usg=AOvVaw2nPmIjOOE1iWk2CByyeClw, accessed 28 December 2017.

8. T. Schlanger and C. B. Philips. "The mutual fund graveyard: An analysis of dead funds," The Vanguard Group, January 2013.

9. https://xkcd.com/1827/.

10. Knowledge Extraction Based on Evolutionary Learning, http://sci2s.ugr.es/keel/dataset.php?cod=163,accessed 22 September 2019.

11. M. C. Bryson, "The *Literary Digest* poll: Making of a statistical myth," *The American Statistician* **30** (1976):184–5.

12. http://www.applied-survey-methods.com/nonresp.html, accessed 4 November 2018.

13. Office for National Statistics, https://www.ons.gov.uk/employmentandlabourmarket/peopleinwork/employmentandemployeetypes/methodologies/labourforcesurveyperformanceandqualitymonitoringreports/labourforcesurveyperformanceandqualitymonitoringreportjulytoseptember2017.

14. R. Tourangeau and T. J. Plewes, eds., *Nonresponse in Social Surveys: A Research Agenda* (Washington, DC: National Academies Press, 2013).

15. J. Leenheer and A. C. Scherpenzeel, "Does it pay off to include non-internet households in an internet panel?" *International Journal of Internet Science* **8** (2013), 17–29.

16. Tourangeau and Plewes, *Nonresponse in Social Surveys.*

17. H. Wainer, "Curbstoning IQ and the 2000 presidential election," *Chance* **17** (2004): 43–46.

18. I. Chalmers, E. Dukan, S. Podolsky, and G. D. Smith, "The advent of fair treatment allocation schedules in clinical trials during the 19th and early 20th centuries," *Journal of the Royal Society of Medicine* **105** (2012): 221–7.

19. J. B. Van Helmont, *Ortus Medicinae, The Dawn of Medicine* (Amsterdam: Apud Ludovicum Elzevirium, 1648), http://www.jameslindlibrary.org/van-helmont-jb-1648/, accessed 15 June 2018.

20. W. W. Busse, P. Chervinsky, J. Condemi, W. R. Lumry, T. L. Petty, S. Rennard, and R. G. Townley, "Budesonide delivered by Turbuhaler is effective in a dose-dependent fashion when used in the treatment of adult patients with chronic

asthma," *Journal of Allergy and Clinical Immunology* **101** (1998): 457–63; J. R. Carpenter and M. Kenward, "Missing data in randomised controlled trials: A practical guide," November 21, 2007, http://citeseerx.ist.psu.edu/viewdoc/download?doi=10.1.1.468.9391&rep=rep1&type=pdf, accessed 7 May 2018.

21. P. K. Robins, "A comparison of the labor supply findings from the four negative income tax experiments," *Journal of Human Resources* **20** (1985): 567–82.

22. A. Leigh, *Randomistas: How Radical Researchers Are Changing Our World* (New Haven, CT: Yale University Press, 2018).

23. P. Quinton, "The impact of information about crime and policing on public perceptions," National Policing Improvement Agency, January 2011, http://whatworks.college.police.uk/Research/Documents/Full_Report_-_Crime_and_Policing_Information.pdf, accessed 17 June 2018.

24. J. E. Berecochea and D. R. Jaman, (1983) *Time Served in Prison and Parole Outcome: An Experimental Study: Report Number 2*, Research Division, California Department of Corrections.

25. G.C.S. Smith and J. Pell, "Parachute use to prevent death and major trauma related to gravitational challenge: Systematic review of randomised controlled trials," *British Medical Journal* **327** (2003): 1459–61.

26. *Washington Post*, "Test of 'dynamic pricing' angers Amazon customers," October 7, 2000, http://www.citi.columbia.edu/B8210/read10/Amazon%20Dynamic%20Pricing%20Angers%20Customers.pdf, accessed 19 June 2018.

27. BBC, "Facebook admits failings over emotion manipulation study," *BBC News*, 3 October 2014, https://www.bbc.co.uk/news/technology-29475019, accessed 19 June 2018.

Chapter 3: Definitions and Dark Data

1. http://www.bbc.co.uk/news/uk-politics-eu-referendum-35959949.

2. Immigration figures, https://www.ons.gov.uk/peoplepopulationandcommunity/populationandmigration/internationalmigration/articles/noteonthedifferencebetweennationalinsurancenumberregistrationsandtheestimateoflongterminternationalmigration/2016, accessed 2 January 2018.

3. Office for National Statistics, "Crime in England and Wales: Year ending June 2017," https://www.ons.gov.uk/peoplepopulationandcommunity/crimeandjustice/bulletins/crimeinenglandandwales/june2017#quality-and-methodology, accessed 4 January 2018.

4. J. Wright, "The real reasons autism rates are up in the U.S." *Scientific American*, March 3, 2017, https://www.scientificamerican.com/article/the-real-reasons-autism-rates-are-up-in-the-u-s/, accessed 3 July 2018.

5. N. Mukadam, G. Livingston, K. Rantell, and S. Rickman, "Diagnostic rates and treatment of dementia before and after launch of a national dementia policy: An

observational study using English national databases. *BMJ Open* **4**, no. 1 (January 2014), http://bmjopen.bmj.com/content/bmjopen/4/1/e004119.full.pdf, accessed 3 July 2018.

6. https://www.ons.gov.uk/businessindustryandtrade/retailindustry/timeseries/j4mc/drsi.

7. https://www.census.gov/retail/mrts/www/data/pdf/ec_current.pdf.

8. Titanic Disaster: Official Casualty Figures, 1997, http://www.anesi.com/titanic.htm, accessed 2 October 2018.

9. A. Agresti, *Categorical Data Analysis*, 2d ed. (New York: Wiley, 2002), 48–51.

10. W. S. Robinson, "Ecological correlations and the behavior of individuals," *American Sociological Review* **15** (1950): 351–7.

11. G. Gigerenzer, *Risk Savvy: How to Make Good Decisions* (London: Penguin Books, 2014), 202.

12. W. J. Krzanowski, *Principles of Multivariate Analysis*, rev. ed. (Oxford: Oxford University Press, 2000), 144.

Chapter 4: Unintentional Dark Data

1. S. de Lusignan, J. Belsey, N. Hague, and B. Dzregah, "End-digit preference in blood pressure recordings of patients with ischaemic heart disease in primary care," *Journal of Human Hypertension* **18** (2004): 261–5.

2. L. E. Ramsay et al., "Guidelines for management of hypertension: Report of the third working party of the British Hypertension Society," *Journal of Human Hypertension* **13** (1999): 569–92.

3. J. M. Roberts Jr. and D. D. Brewer, "Measures and tests of heaping in discrete quantitative distributions," *Journal of Applied Statistics* **28** (2001): 887–96.

4. https://www.healthline.com/health/mens-health/average-weight-for-men.

5. B. Kenber, P. Morgan-Bentley, and L. Goddard, "Drug prices: NHS wastes £30m a year paying too much for unlicensed drugs, *Times* (London), 26 May 2018, https://www.thetimes.co.uk/article/drug-prices-nhs-wastes-30m-a-year-paying-too-much-for-unlicensed-drugs-kv9kr5m8p?shareToken=0e41d3bbd6525068746b7db8f9852a24, accessed 26 May 2018.

6. H. Wainer, "Curbstoning IQ and the 2000 presidential election," *Chance* **17** (2004): 43–46.

7. W. Kruskal, "Statistics in society: Problems unsolved and unformulated," *Journal of the American Statistical Association*, **76**, (1981): 505–15.

8. I have been unable to find a definitive origin for this law. In his 1979 Presidential Address to the Royal Statistical Society, Claus Moser ("Statistics and public policy," *Journal of the Royal Statistical Society, Series A* **143** (1980): 1–32) says it was devised by

the UK's Central Statistical Office. Andrew Ehrenberg quotes it as Twyman's law without attribution in "The teaching of statistics: Corrections and comments," *Journal of the Royal Statistical Society, Series A* **138** (1975): 543–45.

9. T. C. Redman, "Bad data costs the U.S. $3 trillion per year," *Harvard Business Review*, 22 September 2016, https://hbr.org/2016/09/bad-data-costs-the-u-s-3-trillion-per-year, accessed 17 August 2018.

10. ADRN, https://adrn.ac.uk/.

11. https://adrn.ac.uk/media/174470/homlessness.pdf, accessed 24 August 2018.

Chapter 5: Strategic Dark Data

1. https://eur-lex.europa.eu/legal-content/EN/TXT/PDF/?uri=CELEX:32004L0113, accessed 18 February 2019.

2. M. Hurwitz and J. Lee, *Grade Inflation and the Role of Standardized Testing* (Baltimore, MD: Johns Hopkins University Press, forthcoming).

3. R. Blundell, D. A. Green, and W. Jin, "Big historical increase in numbers did not reduce graduates' relative wages," Institute for Fiscal Studies, 18 August 2016, https://www.ifs.org.uk/publications/8426, accessed 23 November 2018.

4. D. Willetts, *A University Education* (Oxford: Oxford University Press, 2017).

5. R. Sylvester, "Schools are cheating with their GCSE results," *Times* (London), 21 August 2018, https://www.thetimes.co.uk/article/schools-are-cheating-with-their-gcse-results-q83s909k6?shareToken=0ce9828e6183e9b37a1454f8f588eaa7, accessed 23 August 2018.

6. "Ambulance service 'lied over response rates,'" *Telegraph* (London), 28 February 2003, http://www.telegraph.co.uk/news/1423338/Ambulance-service-lied-over-response-rates.html, downloaded on 6 October 2018.

7. https://sites.psu.edu/gershcivicissue/2017/03/15/unemployment-and-how-to-manipulate-with-statistics/, accessed 6 October 2018.

8. https://www.heraldscotland.com/news/13147231.Former_police_officers___crime_figures_are_being_massaged_to_look_better_/.

9. J. M. Keynes, *General Theory of Employment Interest and Money* (New York: Harcourt, Brace, 1936).

10. BBC, 1 February 2011, https://www.bbc.co.uk/news/uk-12330078, accessed 18 August 2018.

11. Direct Line Group, 2014, https://www.directlinegroup.com/media/news/brand/2014/11-07-2014b.aspx, accessed 11 April 2014.

12. A. Reurink, "Financial fraud: A literature review," MPIfG Discussion Paper 16/5 (Cologne: Max Planck Institute for the Study of Societies, 2016).

13. R. Caruana, Y. Lou, J. Gehrke, P. Koch, M. Sturm, and N. Elhahad, "Intelligible models for healthcare: predicting pneumonia risk and hospital 30-day readmission,"

Proceedings of the 21st ACM SIGKDD International Conference on Knowledge Discovery and Data Mining, KDD '15, Sydney, Australia, 10–13 August 2015, pp. 1721–30.

14. Board of Governors of the Federal Reserve System, *Report to the Congress on Credit Scoring and Its Effects on the Availability and Affordability of Credit*, August 2007, https://www.federalreserve.gov/boarddocs/RptCongress/creditscore/creditscore.pdf, accessed 18 August 2018.

15. E. Wall, "How car insurance costs have changed," *Telegraph* (London), 21 January 2013, http://www.telegraph.co.uk/finance/personalfinance/insurance/motorinsurance/9815330/How-car-insurance-costs-have-changed-EU-gender-impact.html, accessed 19 August 2018.

Chapter 6: Intentional Dark Data

1. V. Van Vlasselaer, T. Eliassi-Rad, L. Akoglu, M. Snoeck, and B. Baesens, "Gotcha! Network-based fraud detection for social security fraud," *Management Science* **63** (14 July 2016): 3090–3110.

2. B. Baesens, V. van Vlasselaer, and W. Verbet, *Fraud Analytics: Using Descriptive, Predictive, and Social Network Techniques: A Guide to Data Science for Fraud Detection* (Hoboken, NJ: Wiley, 2105), 19.

3. "Crime in England and Wales: Year Ending June 2017," https://www.ons.gov.uk/peoplepopulationandcommunity/crimeandjustice/bulletins/crimeinenglandandwales/june2017, accessed 31 December 2017.

4. D. J. Hand and G. Blunt, "Estimating the iceberg: How much fraud is there in the UK?" *Journal of Financial Transformation* **25**, part 1(2009): 19–29, http://www.capco.com/?q=content/journal-detail&sid=1094.

5. Rates of fraud, identity theft and scams across the 50 states: FTC data," Journalist's Resource, 4 March 2015, https://journalistsresource.org/studies/government/criminal-justice/united-states-rates-fraud-identity-theft-federal-trade-commission, accessed 19 August 2018.

6. B. Whitaker, "Never too young to have your identity stolen," *New York Times*, 27 July 2007, http://www.nytimes.com/2007/07/21/business/21idtheft.html, accessed 3 February 2018.

7. Javelin, 1 February 2017, https://www.javelinstrategy.com/coverage-area/2017-identity-fraud, accessed 3 February 2018.

8. III, "Facts + Statistics: Identity theft and cybercrime,"2016, https://www.iii.org/fact-statistic/facts-statistics-identity-theft-and-cybercrime#, accessed 3 February 2018.

9. DataShield, 14 March 2013, http://datashieldcorp.com/2013/03/14/5-worst-cases-of-identity-theft-ever/, accessed 3 February 2018.

10. A. Reurink; Chapter 5, Note 12.

11. https://www.sec.gov/news/pressrelease/2015–213.html, accessed 30 September 2018.

12. "Accounting scandals: The dozy watchdogs," *Economist*, 11 December 2014, https://www.economist.com/news/briefing/21635978-some-13-years-after-enron-auditors-still-cant-stop-managers-cooking-books-time-some, accessed 7 April 2018.

13. E. Greenwood, *Playing Dead: A Journey through the World of Death Fraud* (New York: Simon and Schuster, 2017).

14. *CBS This Morning*, "Playing a risky game: People who fake death for big money," https://www.cbsnews.com/news/playing-a-risky-game-people-who-fake-death-for-big-money/, accessed 6 April 2018.

15. M. Evans, "British woman who 'faked death in Zanzibar in £140k insurance fraud bid' arrested along with teenage son," *Telegraph* (London), 15 February 2017, https://www.telegraph.co.uk/news/2017/02/15/british-woman-faked-death-zanzibar-140k-insurance-fraud-bid/, accessed 6 April 2018.

16. S. Hickey, "Insurance cheats discover social media is the real pain in the neck," *Guardian* (London), 18 July 2016, https://www.theguardian.com/money/2016/jul/18/insurance-cheats-social-media-whiplash-false-claimants, accessed 4 April 2018.

17. P. Kerr, " 'Ghost Riders' are target of an insurance sting," *New York Times*, 18 August 1993, https://www.nytimes.com/1993/08/18/us/ghost-riders-are-target-of-an-insurance-sting.html, accessed 6 April 2018.

18. FBI (N.A.), "Insurance Fraud," https://www.fbi.gov/stats-services/publications/insurance-fraud, accessed 6 April 2018.

19. E. Crooks, "More than 100 jailed for fake BP oil spill claims," *Financial Times* (London), 15 January 2017, https://www.ft.com/content/6428c082-db1c-11e6–9d7c-be108f1c1dce, accessed 6 April 2018.

20. ABI, "The con's not on—Insurers thwart 2,400 fraudulent insurance claims valued at £25 million every week," Association of British Insurers, 7 July 2017, https://www.abi.org.uk/news/news-articles/2017/07/the-cons-not-on—insurers-thwart-2400-fraudulent-insurance-claims-valued-at-25-million-every-week/, accessed 4 April 2018.

21. "PwC Global Economic Crime Survey: 2016; Adjusting the lens on economic crime," 18 February 2016, https://www.pwc.com/gx/en/economic-crime-survey/pdf/GlobalEconomicCrimeSurvey2016.pdf, accessed 8 April 2018.

Chapter 7: Science and Dark Data

1. J. M. Masson, ed., *The Complete Letters of Sigmund Freud to Wilhelm Fliess* (Cambridge, MA: Belknap Press, 1985), 398.

2. "Frontal lobotomy," *Journal of the American Medical Association* **117** (16 August 1941): 534–35.

3. N. Weiner, *Cybernetics* (Cambridge, MA: MIT Press, 1948).

4. J. B. Moseley et al., "A controlled trial of arthroscopic surgery for osteoarthritis of the knee," *New England Journal of Medicine* **347**, no. 2 (2002): 81–88.

5. J. Kim et al., Association of multivitamin and mineral supplementation and risk of cardiovascular disease: A systematic review and meta-analysis. *Circulation: Cardiovascular Quality and Outcomes* **11** (July 2018), http://circoutcomes.ahajournals.org/content/11/7/e004224, accessed 14 July 2018.

6. J. Byrne, MD, "Medical practices not supported by science," *Skeptical Medicine*, https://sites.google.com/site/skepticalmedicine/medical-practices-unsupported-by-science, accessed 14 July 2018.

7. T. Kuhn, *The Structure of Scientific Revolutions*, 2d ed. (Chicago: University of Chicago Press, 1970), 52.

8. J.P.A. Ioannidis, "Why most published research findings are false," *PLOS Medicine* **2**, no. 8 (2005): 696–701.

9. L. Osherovich, "Hedging against academic risk," *Science-Business eXchange*, 14 April 2011, https://www.gwern.net/docs/statistics/bias/2011-osherovich.pdf, accessed 12 July 2018.

10. M. Baker, "1,500 scientists lift the lid on reproducibility," *Nature* **533** (July 2016): 452–54, https://www.nature.com/news/1-500-scientists-lift-the-lid-on-reproducibility-1.19970, accessed 12 July 2018.

11. C. G. Begley and L. M. Ellis, "Raise standards for preclinical cancer research," *Nature-Comment* **483** (March 2012): 531–33.

12. L. P. Freedman, I. M. Cockburn, and T. S. Simcoe, "The economics of reproducibility in preclinical research," *PLOS Biology*, 9 June 2015, http://journals.plos.org/plosbiology/article?id=10.1371/journal.pbio.1002165, accessed 12 July 2018.

13. B. Nosek et al., "Estimating the reproducibility of psychological science," *Science* **349**, no. 6251 (August 2015): 943–52.

14. https://cirt.gcu.edu/research/publication_presentation/gcujournals/nonsignificant.

15. http://jir.com/index.html.

16. F. C. Fang, R. G. Steen, and A. Casadevall, "Misconduct accounts for the majority of retracted scientific publications," *PNAS* **109** (October 2012): 17028–33.

17. D. G. Smith, J. Clemens, W. Crede, M. Harvey, and E. J. Gracely, "Impact of multiple comparisons in randomized clinical trials," *American Journal of Medicine* **83** (September 1987): 545–50.

18. C. M. Bennett, A. A. Baird, M. B. Miller, and G. L. Wolford, "Neural correlates of interspecies perspective taking in the post-mortem Atlantic Salmon: An argument for proper multiple comparisons correction," *Journal of Serendipitous and Unexpected Results* **1**, no. 1 (2009): 1–5, http://docplayer.net/5469627-Journal-of-serendipitous-and-unexpected-results.html, accessed 16 August 2018.

19. S. Della Sala and R. Cubelli, "Alleged 'sonic attack' supported by poor neuropsychology," *Cortex* **103** (2018): 387–88.

20. R. L. Swanson et al., "Neurological manifestations among U.S. Government personnel reporting directional audible and sensory phenomena in Havana, Cuba," *JAMA* **319** (20 March 2018): 1125–33.

21. F. Miele, *Intelligence, Race, and Genetics: Conversations with Arthur R. Jensen* (Oxford: Westview Press, 2002), 99–103.

22. C. Babbage, *Reflections on the Decline of Science in England, and on Some of Its Causes* (London: B. Fellowes, 1830).

23. A. D. Sokal, "Transgressing the boundaries: Toward a transformative hermeneutics of quantum gravity," *Social Text* **46/47** (Spring/Summer 1996): 217–52.

24. https://read.dukeupress.edu/social-text, accessed 23 January 2019.

25. A. Sokal and J. Bricmont, *Intellectual Imposters: Postmodern Philosophers' Abuse of Science* (London: Profile Books, 1998).

26. http://science.sciencemag.org/content/342/6154/60/tab-pdf.

27. http://www.scs.stanford.edu/~dm/home/papers/remove.pdf.

28. https://j4mb.org.uk/2019/01/09/peter-boghossian-professor-faces-sack-over-hoax-that-fooled-academic-journals/.

29. C. Dawson and A. Smith Woodward, "On a bone implement from Piltdown (Sussex)," *Geological Magazine* **Decade 6**, no. 2 (1915): 1–5, http://www.boneandstone.com/articles_classics/dawson_04.pdf, accessed 7 July 2018.

30. M. Russell (2003) *Piltdown Man: The Secret Life of Charles Dawson* (Stroud, UK: Tempus, 2003); M. Russell, *The Piltdown Man Hoax: Case Closed* (Stroud, UK: The History Press, 2012).

31. J. Scott, "At UC San Diego: Unraveling a research fraud case," *Los Angeles Times*, 30 April 1987, http://articles.latimes.com/1987-04-30/news/mn-2837_1_uc-san-diego, accessed 4 July 2018.

32. B. Grant, "Peer-review fraud scheme uncovered in China," *Scientist*, 31 July 2017, https://www.the-scientist.com/the-nutshell/peer-review-fraud-scheme-uncovered-in-china-31152, accessed 4 July 2018.

33. https://ori.hhs.gov/about-ori, accessed 14 October 2018.

34. R. A. Millikan, "On the elementary electric charge and the Avogadro constant," *Physical Review* **2**, no. 2 (August 1913): 109–43.

35. W. Broad and N. Wade, *Betrayers of the Truth: Fraud and Deceit in the Halls of Science* (New York: Touchstone, 1982).

36. D. Goodstein, "In defense of Robert Andrews Millikan," *American Scientist* **89**, no. 1 (January-February 2001): 54–60.

37. R. G. Steen, A. Casadevall, and F. C. Fang, "Why has the number of scientific retractions increased?" *PLOS ONE* **8**, no. 7 (8 July 2013), http://journals.plos.org/plosone/article?id=10.1371/journal.pone.0068397, accessed 9 July 2018.

38. D. J. Hand, "Deception and dishonesty with data: Fraud in science," *Significance* **4**, no.1 (2007): 22–25; D. J. Hand, *Information Generation: How Data Rule Our World* (London: Oneworld Publications, 2007); H. F. Judson, *The Great Betrayal: Fraud in Science* (Orlando, FL: Harcourt, 2004).

39. D. J. Hand, "Who told you that?: Data provenance, false facts, and separating the liars from the truth-tellers," *Significance* (August 2018): 8–9.

40. LGTC (2015), https://assets.publishing.service.gov.uk/government/uploads/system/uploads/attachment_data/file/408386/150227_PUBLICATION_Final_LGTC_2015.pdf, accessed 17 April 2018.

41. Tameside, https://www.tameside.gov.uk/Legal/Transparency-in-Local-Government, accessed 17 April 2018.

Chapter 8: Dealing with Dark Data

1. See, for example, D. Rubin, "Inference and missing data," *Biometrika*, **63**, no. 3 (December 1976): 581–92.

2. C. Marsh, *Exploring Data* (Cambridge: Cambridge University Press, 1988).

3. X.-L. Meng, "Statistical paradises and paradoxes in big data (I): Law of large populations, big data paradox, and the 2016 U.S. presidential election,"*Annals of Applied Statistics* **12** (June 2018): 685–726.

4. R.J.A. Little, "A test of missing completely at random for multivariate data with missing values," *Journal of the American Statistical Association* **83**, no. 404 (December 1988): 1198–1202.

5. E. L. Kaplan and P. Meier, "Nonparametric estimation from incomplete observations," *Journal of the American Statistical Association* **53**, no. 282 (June 1958): 457–81.

6. G. Dvorsky, "What are the most cited research papers of all time?" 30 October 2014, https://io9.gizmodo.com/what-are-the-most-cited-research-papers-of-all-time-1652707091, accessed 22 April 2018.

7. F. J. Molnar, B. Hutton, and D. Fergusson, "Does analysis using 'last observation carried forward' introduce bias in dementia research?" *Canadian Medical Association Journal* **179** no. 8 (October 2008):751–53.

8. J. M. Lachin, "Fallacies of last observation carried forward," *Clinical Trials* **13**, no. 2 (April 2016): 161–68.

9. A. Karahalios, L. Baglietto, J. B. Carlin, D. R. English, and J. A. Simpson, "A review of the reporting and handling of missing data in cohort studies with repeated assessment of exposure measures," *BMC Medical Research Methodology* **12** (11 July 2012): 96, https://bmcmedresmethodol.biomedcentral.com/track/pdf/10.1186/1471-2288-12-96.

10. S.J.W. Shoop, "Should we ban the use of 'last observation carried forward' analysis in epidemiological studies?" *SM Journal of Public Health and Epidemiology* **1**, no. 1(June 2015): 1004.

11. S. J. Miller, ed., *Benford's Law: Theory and Applications* (Princeton, NJ: Princeton University Press, 2015).

Chapter 9: Benefitting from Dark Data

1. S. Newcomb "Measures of the velocity of light made under the direction of the Secretary of the Navy during the years 1880–1882," *Astronomical Papers* **2** (1891): 107–230 (Washington, DC: U.S. Nautical Almanac Office).

2. ADRN, https://adrn.ac.uk/.

3. D. Barth-Jones D. "The 're-identification' of Governor William Weld's medical information: A critical re-examination of health data identification risks and privacy protections, then and now," 3 September 2015, https://papers.ssrn.com/sol3/papers.cfm?abstract_id=2076397, accessed 24 June 2018.

4. A. Narayanan and V. Shmatikov, "How to break the anonymity of the Netflix Prize dataset," 22 November 2007, https://arxiv.org/abs/cs/0610105, accessed 25 March 2018; A. Narayanan and V. Shmatikov V. (2008) Robust de-anonymization of large sparse datasets (how to break the anonymity of the Netflix Prize dataset), 5 February 2008, https://arxiv.org/pdf/cs/0610105.pdf, accessed 24 June 2018.

5. D. Hugh-Jones, "Honesty and beliefs about honesty in 15 countries," 29 October 2015, https://www.uea.ac.uk/documents/3154295/7054672/Honesty+paper/41fecf09–235e-45c1-afc2-b872ea0ac882, accessed 26 June 2018.

6. C. Gentry, "Computing arbitrary functions of encrypted data," *Communications of the ACM*, **53**, no. 3 (March 2010): 97–105.

Chapter 10: Classifying Dark Data

1. https://www.behaviouralinsights.co.uk/wp-content/uploads/2016/08/16-07-12-Counting-Calories-Final.pdf, accessed 27 October 2018.

2. A. Cavallo, "Online and official price indexes: Measuring Argentina's inflation," *Journal of Monetary Economics* **60**, no. 2 (2013): 152–65.

3. A. Cavallo and R. Rigobon, "The billion prices project: Using online prices for measurement and research," *Journal of Economic Perspectives* **30**, no. 2 (Spring 2016): 151–78.

4. C. Szegedy et al., "Intriguing properties of neural networks," https://arxiv.org/pdf/1312.6199.pdf, 19 February 2014, accessed 23 August 2008.

5. M. Sharif, S. Bhagavatula, L. Bauer, and M. K. Reiter, "Accessorize to a crime: Real and stealthy attacks on state-of-the-art face recognition," October 2016, https://www.cs.cmu.edu/~sbhagava/papers/face-rec-ccs16.pdf, accessed 23 August 2018.

INDEX

A NOTE ON THE TYPE

This book has been composed in Arno, an Old-style serif typeface in the classic Venetian tradition, designed by Robert Slimbach at Adobe.

CPSIA information can be obtained
at www.ICGtesting.com
Printed in the USA
JSHW040926120122
21950JS00002B/2